Synthese Library

Studies in Epistemology, Logic, Methodology, and Philosophy of Science

Volume 377

More information about this series at http://www.springer.com/series/6607

Harmen Ghijsen

The Puzzle of Perceptual Justification

Conscious experience, Higher-order Beliefs, and Reliable Processes

Springer

Harmen Ghijsen
Centre for Logic and Analytic Philosophy
Institute of Philosophy, KU Leuven
Leuven, Belgium

Synthese Library
ISBN 978-3-319-30498-4 ISBN 978-3-319-30500-4 (eBook)
DOI 10.1007/978-3-319-30500-4

Library of Congress Control Number: 2016937332

Printed on acid-free paper

This Springer imprint is published by Springer Nature
The registered company is Springer International Publishing AG Switzerland

Preface

We're all probably familiar with certain common perceptual illusions: the black tie bought in the shop turns out to be dark blue, the monster in the bedroom is just the shadow of a tree in the backyard, and no one actually called your name even though it sounded just like it. Although you would probably make a false judgment in these circumstances, you appear to have a good reason for it: you consciously experienced the world as being that way. That is, even though you did not *know* that such-and-so was the case (because you were actually wrong), at least you were *justified* in your belief.

This notion of *perceptual justification*, and its relation to conscious experience and reasons, will be the main topic of investigation in this book. Let me note immediately that I do not want to commit myself to there being different kinds of justification (perceptual, memorial, inferential, etc.). I just want to focus on the justification that typically arises in successful cases of perception, whether or not this justification is of a special kind. I take it that justification is a property which gives beliefs a certain kind of positive epistemic value while remaining weaker than knowledge: a justified belief is one which is epistemically better than an unjustified belief even if for some reason it does not amount to knowledge. I'm not too concerned with whether "justification" is the right term for this, perhaps some would prefer "entitlement" or "warrant" or some other word. What I'm investigating is simply that property which typically gives perceptual beliefs their positive epistemic status, and I'll refer to that property with the term "justification."

I take this to be an interesting notion to investigate for several reasons. First of all, perception is one of our fundamental ways of gaining knowledge about the world. Given that many people take justification to be a necessary condition for knowledge, an investigation of perceptual justification could help to better our understanding of the fundamental source of knowledge that perception is. Furthermore, an investigation of perceptual justification helps to shed light on the nature of justification (and knowledge) in general, given the important role perception plays in the acquisition of many of our beliefs. If a theory of justification is unsuccessful for the case of perception, then it can be a radically incomplete theory at best. At worst, it simply is completely mistaken.

But not only is perception one of *our* fundamental sources of knowledge, it also appears to be a fundamental source of knowledge for less cognitively sophisticated epistemic agents, such as small children and animals. It would be nice to have a theory of justification that could secure the continuity between these unsophisticated cognizers and ourselves while also allowing for some important differences because of the level of cognitive sophistication. Such a strategy seems important not only for epistemology, but also for, say, philosophy of mind and action theory. If one focuses too much on the peculiarities of how we, adult human agents, are related to the relevant *analysandum*, then one's overall theory is likely to become too demanding to work for unsophisticated cognizers. This might not always be a bad result, but at least for the case of epistemology, this looks like a problem.

The externalist view of perceptual justification that I aim to defend is one that seeks to accommodate the possibility of animal knowledge, a virtue that has commonly been stressed as a motivation for externalist views in general. But it also attempts to make room for the peculiarities of human belief formation in acknowledgement of the fact that there are some important differences between sophisticated and unsophisticated agents. Not only does such a view combine two elements that are prima facie desirable in any epistemological theory, it will also help to solve some classic problems for externalism and some general problems for any theory of (perceptual) justification.

One important aspect of this approach has to do with distinguishing between evidential and non-evidential justification. Sometimes we come to justifiably believe new things on the basis of other things we know; this latter knowledge would then act as evidence for our new beliefs, making the cases into instances of evidential justification. For instance, I can come to justifiably believe that you won't be on time for dinner by reflecting on the fact that you told me that you had a meeting at 6 PM and the fact that your meetings tend to take a long time. However, there might also be instances in which I come to have justified beliefs without these beliefs being based on any evidence, which are instances of non-evidential justification. The main thesis of this book is that perceptual justification is best construed as such a form of non-evidential justification: whenever we perceptually experience that something is the case, we (normally) just thereby also believe that it is the case, without having to base those beliefs on the relevant perceptual experiences. In such a case, the justifier of the belief is a non-evidential one, namely, the reliability of the perceptual process. Of course, when challenged, we might supply additional *evidential* justifiers for our perceptual beliefs. For instance, if someone challenges my claim to know that there is some milk left in the fridge, I could respond by saying that I *saw* that there was some milk left in the fridge. The important point, though, is that even if we can cite and use this additional evidence to justify our perceptual beliefs, this does not mean that we needed the additional evidence to make our beliefs justified in the first place. It does mean that we, sophisticated cognizers, have justifiers at our disposal that are not available to less sophisticated ones.

This latter aspect is related to what is distinctive for sophisticated cognizers: the capability for higher-order thought. Not only do we often see, and thereby know, that such-and-so is the case, we normally also know *that we are seeing* that such-and-so is the case. And for this type of higher-order knowledge, a similar question arises as before: is this knowledge best analyzed as depending on evidential justification on the basis of experience, or is it better seen as depending on non-evidential justification? Again, the preferable answer is the latter one. It is a mistake to think that conscious experience itself provides evidence on the basis of which we conclude that we are currently seeing that such-and-so is the case, a mistake that lies at the bottom of some persistent philosophical problems. Once one accepts the possibility of non-evidential justification for first-order and higher-order beliefs, one can answer these philosophical problems and present a persuading externalist account of perceptual justification.

This externalist view does limit the role of conscious experience in epistemology. In effect, the view argues that reliability is far more important for perceptual justification than conscious experience, something that seems to go against some of our prima facie intuitions about justification. That is why I start with a critical discussion of contemporary theories of perceptual justification that argue in favor of the idea of experiential evidence and leave discussion of my own preferred externalist alternative until later.

The precise outline of the book will be as follows. Chapter 1 introduces the theories of perceptual justification to be discussed in relation to a challenge that arises from the indistinguishability of perception and hallucination. Each of these theories will highlight different aspects of perception, namely, conscious experience, higher-order belief, and reliability of the perceptual process.

Chapters 2 and 3 are both devoted to *experientialist* views of perceptual justification, which hold that perceptual experiences justify perceptual beliefs by acting as their evidence. Chapter 2 is devoted to variants of *evidentialism*, which hold that perceptual experience can fulfill its evidential role without having propositional content. Chapter 3 is devoted to variants of *dogmatism*, which hold that a perceptual experience with the propositional content that p is sufficient for immediate (prima facie) justification of the belief that p. The overall problem for these experientialist views will be presented in the form of a (Sellarsian) dilemma: if a perceptual experience lacks propositional content, then it is entirely unclear how it can serve as evidence for belief, but if a perceptual experience has propositional content, then one has to explain how it is able to do so without being justified itself.

Chapter 4 discusses variants of *epistemological disjunctivism*. According to these accounts, perceptual justification has to do with having access to factive reasons of the form "I see that p." Epistemological disjunctivism thus holds, in agreement with experientialism, that perceptual justification has to do with having evidence, but in contrast with experientialism, it takes this evidence to consist in factive reasons. The largest problem I present for this view is that of hyper-intellectualization: having access to factive reasons plausibly requires having the capacity for higher-order beliefs, which is too cognitively demanding for unsophisticated epistemic agents. This chapter thus not only critically discusses an alternative account of evidential perceptual justification, it also starts the discussion of the role of higher-order beliefs in perceptual justification.

After displaying the problems of several accounts of perceptual justification that connect justification to evidence, Chap. 5 introduces a non-evidential view of justification: *process reliabilism*. The first part of this chapter focuses on the classic account of process reliabilism, which holds that the reliability of a specific type of belief-forming process determines whether a belief is justified. The second part discusses two alternatives to this classic account, i.e., inferentialist reliabilism and proper functionalism, in the light of the well-known New Evil Demon Problem (which argues against the necessity of reliability for justification) and Clairvoyance Problem (which argues against the sufficiency of reliability for justification).

In Chap. 6 I integrate several insights from epistemological disjunctivism, inferentialist reliabilism, and proper functionalism to account for the Clairvoyance Problem and New Evil Demon Problem. Although Chap. 4 shows that higher-order beliefs should not be taken as necessary for perceptual justification, they can still play a role in providing additional evidential justification for perceptual beliefs as long as they are the output of reliable introspective mechanisms. What's more, the fact that we have these introspective mechanisms can also be used to explain, first, how perceptual beliefs get defeated in cases of clairvoyance and, second, why we would overestimate the importance of experience in providing perceptual justification—thereby leading to the mistaken New Evil Demon Intuition. With the Clairvoyance and New Evil Demon Problem out of the way, I conclude that a non-evidential theory of perceptual justification definitely comes out on top.

Leuven, Belgium Harmen Ghijsen

Acknowledgments

This book would probably not have been written in this form if it had not been for a number of people. A lot of the research carried out for this book happened by means of a FLOF doctoral fellowship and a BOF PDMK postdoctoral fellowship at KU Leuven, where I was originally hired as a doctoral researcher by Stefaan Cuypers. He has been a major support ever since, and I have had his full confidence from the moment I started. Chris Kelp has definitely had the most impact on the content of this book, as some parts of it are even based on our joint work. He has influenced my thinking on the issues in this book from the moment we met and has always made time to discuss and comment on my work even before he could have gained any benefit from that himself. Jack Lyons has, unknowingly at first, also been an influence on my writings, and readers familiar with his *Perception and Basic Beliefs* will surely recognize some of the arguments presented here. I would also like to thank all members of the Leuven Epistemology Group, especially Jan Heylen, Mona Simion, and Fernando Broncano-Berrocal, for their valuable input on parts of the book. The same goes for my other colleagues at the Centre for Logic and Analytic Philosophy and all other people that have given me good comments and suggestions at presentations of my work.

On the personal front, I have also been supported by lots of friends and family, some of which I am also fortunate enough to have already mentioned as coworkers. Most importantly here, I'd like to thank my wife, Lorraine Fliek, for her utmost confidence in my abilities.

Parts of this book are based on some of my earlier published work. Below I specify these parts in more detail.

- Sections 2.4, 2.5, and 5.5 are partly based on points made in "The Non-Evidential Nature of Perceptual Experience," *Logique et Analyse* 57 (2014).
- Sections 3.3 and 3.5 are based on "Grounding Perceptual Dogmatism: What are Perceptual Seemings?" *The Southern Journal of Philosophy* 53 (2015): 196–215.

- Sections 3.2, 3.4, and 6.3.1 are based on "Phenomenalist Dogmatist Experientialism and the Distinctiveness Problem," *Synthese* 191 (2014): 1549–1566. © Springer. Reprinted with permission.
- Sections 4.2, 4.3, 4.4 and 4.5 are almost entirely based on a joint paper with Christoph Kelp, "Perceptual Justification: Factive Reasons and Fallible Virtues," in eds. Chienkuo Mi, Michael Slote, and Ernest Sosa, *Moral and Intellectual Virtues*, pp. 164–183 (2016), Routledge. © Taylor and Francis Groups LLC books. Reprinted with permission. The last part of this paper, on knowledge-first virtue epistemology, has not been included in this book, as that more properly belongs to Chris Kelp's own work rather than our joint work.
- Section 4.3.1 is based on "The Basis Problem for Epistemological Disjunctivism Revisited," *Erkenntnis* 80 (2015): 1147–1156. © Springer. Reprinted with permission.
- Sections 5.5, 5.6, and 6.4 are partly based on "Norman and Truetemp Revisited Reliabilistically: A Proper Functionalist Defeat Account of Clairvoyance," *Episteme* 13 (2016): 89–110. © Cambridge University Press. Reprinted with permission.

Contents

Chapter 1
Perception, Hallucination and Justification

1.1 Introduction

In this first chapter, I briefly introduce the three main theories of perceptual justification to be discussed in the following chapters: experientialism, epistemological disjunctivism, and process reliabilism. In Sect. 1.2, I start by considering our ordinary take on perception as one of our fundamental sources of justification, and the way this ordinary conception is challenged by the possibility of illusion and hallucination. I then discuss how experience might be taken to provide a response to this challenge according to sense-datum theorists, evidentialists, and dogmatists about perception. Although the sense-datum theory of perception is by now deemed highly problematic by most philosophers, some aspect of it, i.e., the special epistemic status of experience, is retained in experientialist theories of perceptual justification.

In Sect. 1.3, I discuss a rather different way to respond to the challenge that arises from cases of illusion and hallucination. Epistemological disjunctivism maintains that one's evidence is crucially different in cases of perception and hallucination even though these cases are introspectively indistinguishable. The version of epistemological disjunctivism that will be of most interest to us here adds that this evidence is constituted not by experience itself, but by the fact that one sees that such-and-so is the case. The most problematic aspect of the theory then consists in explaining how this fact can be accessed.

In Sect. 1.4, I discuss a final way in which perceptual justification can be construed. According to process reliabilism, what matters for perceptual justification is that one's perceptual beliefs have been produced by a specific sort of reliable process. So even if perception and hallucination are introspectively indistinguishable, that does not mean that they provide the same evidence. In fact, on a process reliabilist account, perceptual justification might not have anything to do with evidence at all. One might worry that such an account will make perceptual

H. Ghijsen, *The Puzzle of Perceptual Justification*, Synthese Library 377,
DOI 10.1007/978-3-319-30500-4_1

justification too unsophisticated, given the capacities we have for asking and giving reasons for beliefs. However, process reliabilism is far more permissible with regard to such types of justification than might at first be supposed.

Section 1.5 will briefly sum up the results of this introduction, and outlines the work to be done in the upcoming chapters.

1.2 Experientialism

Perception is not just one of our fundamental sources of knowledge, it also constitutes one of our fundamental sources of *justification*. We can and do easily justify many of our beliefs about the environment by invoking our senses in some way. For instance, if someone asks you why you believe that dinner is ready, you can respond by saying that you heard the kitchen timer go off, or that you can smell that the lasagna is done. Or if someone wants to know what reason you have for believing that person X is at the party, you can respond by saying that you can see person X standing over there. Now, for the moment, we need not worry about whether all these cases of justification are relevantly alike; at this moment we merely need to register the fact that perception does appear to play an important role in our ordinary take on justification as something to appeal to when our claims are challenged.

However, this ordinary take on justification is threatened by considerations regarding the fallibility of our perceptual capacities. We are all familiar with certain common perceptual illusions in which things appear other as they really are. You might misread a sentence and only get it right after someone else points out the mistake, or you might come to realize the actual color of your shirt when you are no longer at the store that sold it to you. In such cases, you seem to form your beliefs on the basis of perception in exactly the same way as you do in other circumstances, yet you end up with false beliefs about the environment. Now this in itself does not seem to pose too much of a problem for the idea that perception can provide us with justification for our beliefs. The mentioned perceptual mistakes are all relatively local in the sense that they are about just one property of the environment, and they are also relatively rare in the sense that we are still generally reliable with regard to the properties we happen to be mistaken about in this one case. Indeed, one might even want to claim that we still have justified beliefs in the above cases of illusion *because* perception is still broadly accurate and reliable, thereby making it epistemically appropriate to base one's beliefs on perception.

Even though local cases of illusion thus do not seem to threaten our ordinary take on perceptual justification, one can extrapolate from them to cases of *massive* deception. In such cases, *all* of our beliefs turn out to be false, even though they appear to be formed in the same way as when they turn out to be true. For instance, one might unknowingly undergo perceptual illusions all the time, or one might even be having a full blown hallucination and on that basis come to have all sorts of false beliefs about the environment. If this were the case, then from your perspective you would still appear to be doing what you were always doing when you formed beliefs on the basis of perception. Nevertheless, all of your beliefs would turn out to be false. Reflection on these sorts of cases has led many philosophers, most notably sense-datum theorists, to a certain line of thinking about perception that is expressed by H.H. Price as follows:

> When I see a tomato there is much that I can doubt. I can doubt whether it is a tomato that I am seeing, and not a cleverly painted piece of wax. I can doubt whether there is any material thing there at all. Perhaps what I took for a tomato was really a reflection; perhaps I am even the victim of some hallucination. One thing however I cannot doubt: that there exists a red patch of a round and somewhat bulgy shape, standing out from a background of other colour-patches, and having a certain visual depth, and that this whole field of colour is directly present to my consciousness.
>
> (Price 1932, p. 3)

The sense-datum theorist's motivation for doubting ordinary cases of perception stems from the mentioned possibilities of non-veridical cases of perception or hallucination in which things nevertheless appear exactly the same. According to the sense-datum theorist, even in those bad cases there is something about which we cannot really be in any doubt, and that is the fact that one is experiencing a certain type of *sense-datum*, such as a red patch of a round and somewhat bulgy shape. Because one cannot be in any doubt about such facts about one's own experiences, sense-datum theorists take beliefs about these facts to be the proper epistemic starting points. From these beliefs one can then infer beliefs about one's environment.

If the sense-datum theorist is correct, then our ordinary take on perception and justification appears to be in need of some refinement. Although we commonly use beliefs about what we perceive when someone challenges our justification for our claims, these perceptual beliefs should not be taken as the epistemic foundation for our other beliefs. In fact, if the sense datum theorist is correct, then it seems that perception is not really a source of justification after all. It's *introspection* that delivers the epistemic foundation for belief, and from there we can then *infer* other beliefs about the world.

Underlying this line of thinking are two ideas that are part of the *classical foundationalist* view of justification. First, there is the idea that we are infallible with regard to our own experiences: we cannot seriously doubt or be mistaken about the fact that a certain sort of red sense-datum is present to consciousness when we seem to see a tomato. Whether we are actually perceiving a tomato or merely hallucinating it, in both cases we can be sure of the fact that we are having a specific sort of experience. Second, there is the idea that we are *more* certain about our own experiences than we are about objects in the external world: even though we cannot doubt that a certain sense-datum is present to consciousness, we *can* doubt whether we are actually seeing a tomato when we seem to see it.

On this classic version of foundationalism, beliefs about our own experiences are epistemically *basic* because they are infallible, and beliefs about the external world can be justifiably inferred from these basic beliefs. In contrast, many contemporary philosophers would reject this version of foundationalism because they no longer accept the requirement of infallibilism for (the foundation of our) justification, nor the putative fact that we are infallible about our own experiences. Most philosophers nowadays agree, in part because of empirical results from cognitive science, that we can even make mistakes about our own experiences (e.g., Dennett 1991; Hurlburt and Schwitzgebel 2007; Bayne and Spener 2010). Thus, instead of starting from such fallible beliefs about our own experiences, contemporary foundationalists simply take fallible *perceptual* beliefs (e.g., that there is a tomato in front of you) as epistemically basic. That is, contemporary foundationalists take perceptual beliefs to be justified without relying on other beliefs for that justification. This has the added bonus of fitting our perceptual phenomenology in which we are usually never aware of *inferring* beliefs about the world from beliefs about our own experiences. What's more, it also prevents the need of finding a way to ground such dubious inferences, as the gap between beliefs about sense-data and beliefs about worldly objects seems too large to overcome without controversial additional premises.

Despite these differences between the classic and contemporary versions of foundationalism, some aspect of classical foundationalism is often retained in contemporary discussions about perception. This aspect has to do with the importance of experiences as *evidence* for our beliefs about the world. The idea is that even if we cannot, and need not, justifiably infer beliefs about the world from beliefs about our own experiences, experience could still play an important role in *justifying* our beliefs about the world. On this account, experience would retain an epistemically privileged position to belief: experience would provide the epistemic foundation for basic beliefs without being in need of such a foundation itself. Our perceptual beliefs would then still be justified without relying on other *beliefs* for their justification, and so they would still be epistemically basic. However, their justificatory source would, on this account, stem from experiences that are in no need of justification themselves. Following Jack Lyons (2009), I'll label this kind of foundationalist account as *experientialism*.

Experientialism has been a popular view among *internalists* about justification. These internalists hold that justification has to do with having a sufficient amount of evidence that is either *accessible* to the subject (on the version of internalism developed in e.g., (Chisholm 1977; BonJour 1985; Steup 1999), usually called *access internalism*) or internal to the subject's mind (on the version developed by Feldman and Conee (2001), called *mentalism*). It's easy to see why internalists of either variety would be attracted to experientialism: experience seems to be the kind of thing to which we always have some sort of access, and it is definitely an important part of our ongoing mental life.

However, some of the problems of sense-datum theorists carry over to the experientialist account as well. If having a certain experience leaves open whether one is genuinely perceiving, under some sort of perceptual illusion, or in the grips of a full blown hallucination, then one might doubt that experiences can really justify perceptual beliefs. Even though contemporary experientialists no longer require an inference on the part of the subject from a belief about one's experience to a belief about one's current environment, they still think of experiences as providing evidence for external world beliefs. So even though experientialists do not have the problem of explaining why inferences from experiences to the world are epistemically permissible, they still have to explain how experiences can *justify* beliefs about the world, given that the experiences are compatible with being in cases of illusion or hallucination.

Earl Conee and Richard Feldman, the most prominent defenders of *evidentialism*—a view according to which one has justification for a belief that *p* if and only if the belief that *p* fits one's evidence—consider this worry about experience and respond as follows:

> [A] possibility that is allowed by one's basic evidence, and incompatible with P, is not automatically a reason to doubt P. The possibility must have an epistemic status that bears on how things actually are, such as evidence that the possibility obtains, or at least evidence that gives the possibility some positive probability of being true [...] Though this is not obvious, reason to doubt P is evidence that a possibility contrary to P obtains in fact. Evidence of sheer possibility is not that.
>
> (Conee and Feldman 2004b, p. 301)

Conee and Feldman here can be taking to argue that, even though having certain perceptual experiences is compatible with being in a case of illusion or hallucination, this does not preclude the experiences from justifying one's beliefs about the world. To reach this conclusion, they seem to construct the open possibility of illusion and hallucination as a *defeater* for one's beliefs about the world. On their view, an experience of, say, a tomato, is good evidence that there is a tomato before you, and the open possibility that you are now having an illusion or hallucination of a tomato can only defeat the justification provided by this evidence if there is positive evidence that this possibility actually obtains. For instance, the situation would be different if a trustworthy source had informed you that you had taken some hallucinatory drug. But given that there normally is no such positive evidence available for the obtaining of cases of illusion or hallucination, the experiential evidence is normally sufficient to justify one's perceptual beliefs. The plausible thought underlying this line of reasoning is that evidence need not entail a proposition to justify a belief in that proposition.

The problem with this response is that the challenge for experientialists can be construed in a way that has nothing to do with either defeat or the fact that evidence need not entail a proposition to justify a belief in that proposition. The challenge for experientialists is to explain why experiences would provide evidence for beliefs about the world at all, rather than evidence for beliefs about having illusions or hallucinations of the world. Why couldn't an experience as of *p* be taken as evidence for the fact that one is currently hallucinating that *p* rather than evidence for the fact that *p* is actually the case? That one is hallucinating that *p* is an alternative explanation of one's having an experience that *p*, and no clear reason has been provided to prefer the explanation that one is perceiving that *p*. So why think that in the normal case a perceptual experience as of *p* is evidence for the proposition that *p* instead of evidence for the proposition that one is having an illusion or hallucination that *p*?

Conee and Feldman also consider this rejoinder on behalf of the adversaries of experientialism, and concede the following:

> [S]ome account is needed of what makes experiences provide basic evidence for ordinary beliefs. It seems clear that experiences do this. Explaining how they do so is a difficult project that remains to be accomplished.
>
> (Conee and Feldman 2004b, p. 303)

So Conee and Feldman agree that explaining why experiences provide evidence for ordinary beliefs about the world is a challenge for their theory, but they do take it as intuitively obvious that experiences are able to do so. I agree that it does seem intuitive that experiences provide evidence for ordinary beliefs, but unfortunately I also think that there are good reasons that count against this intuitive idea. Without an additional answer to the challenge, experientialism will be in trouble.

Conee and Feldman have some hope that *dogmatism*, the view promoted by, e.g., James Pryor (2000) and Michael Huemer (2001), will be able to provide this additional answer. According to dogmatism, a perceptual experience that *p* provides immediate prima facie justification for the belief that *p*. What's more, dogmatists often take the *phenomenology* of perceptual experience to be crucial in explaining why they are able to provide this immediate justification. Perceptual experiences present their content as being actualized here and now, and seem to put you into contact with the truthmakers for this content. But even if this is true, then it still is not clear why such phenomenological facts would make experiences into justifiers for external world beliefs, especially given that hallucinations and illusions have the same phenomenologies.

Both dogmatism and evidentialism thus share some worries in virtue of their common acceptance of experientialism. But there also is an important difference between these two theories. The proponents of dogmatism commonly require that experiences have propositional contents which link them to the beliefs they immediately justify, whereas Conee and Feldman take this requirement to be somewhat unmotivated (Conee and Feldman 2004a, p. 2). Conee and Feldman's evidentialism will thus have more difficulties in explaining which beliefs 'fit' the experiential evidence, even though they can also appeal to experience's non-propositional nature to explain why they have a special epistemic status. I will later argue against these two versions of experientialism on the basis of a Sellarsian dilemma: if experience is non-propositional, as evidentialism has it, then it's unclear how they can serve as evidence for belief; but if experience is propositional, as dogmatists maintain, then it's ad hoc to hold that it can serve as evidence for belief without being justified itself. The first horn of this dilemma will be discussed in combination with evidentialism in Chap. 2, while the second horn will be discussed in combination with dogmatism in Chap. 3.

1.3 Epistemological Disjunctivism

In the previous section we've seen that experientialism inherits some of the problems that plagued sense-datum theory. Where sense-datum theory had problems in accounting for the legitimacy of the inference from beliefs about experience to beliefs about the world, contemporary experientialists have a problem in explaining how perceptual experiences can serve as justifying evidence for beliefs about the world. One of the crucial ingredients for this problem seems to be a thesis about the sameness of evidence in the good cases of perception and the bad cases of illusion and hallucination. As is already implicit in the earlier quote, Price holds that a subject's experience can be the same whether he is genuinely perceiving a tomato or just hallucinating it.[1] His reason for this lies in the "qualitatively indistinguishable" (Price 1932, p. 31) nature of what a subject is confronted with in the case of perception and hallucination. Put somewhat differently, since a subject cannot distinguish between a perceptual experience of a tomato and a hallucinatory experience of a tomato, these experiences must be in an important sense the same.[2] This line of argument, that starts with the indistinguishability of perception and hallucination, and concludes from this that they are of the same kind, is usually known as the argument from hallucination.[3]

Some have argued, contrary to the conclusion of this argument from hallucination, that perception and hallucination are experiences of a radically different kind (Martin 2006; Fish 2009). These *metaphysical disjunctivists* usually stress that indistinguishability does not entail ontological similarity, as it surely is possible that we are unable to distinguish between two *different* things because of our limited powers of discrimination.[4] Sometimes the threat of skepticism is even used to motivate metaphysical disjunctivism: the idea is that one can prevent skepticism from occurring if one does not accept that perception and hallucination are experiences of the same kind (Fish 2009, pp. 24–26).[5] This kind of strategy is

---

[1]Note that this reasoning makes use of the idea that two experiences can be the same. If we think of experiences as unrepeatable events, tied to a specific subject at a specific place and time, then this should be explained more fully. What I will mean by saying that two experiences are the same is that they have the same phenomenal character, where the phenomenal character is understood as what it is like to have the experience (Nagel 1974). As Williamson puts it: "[phenomenal] characters stand to experiences as types of tokens, with respect to a certain mode of classification" (Williamson 2013, p. 49).

[2]Note that Price would not put his point in this way himself, given that he is more concerned with the qualitative indistinguishability of the objects of experiences, i.e., sense-data.

[3]Presentations of this argument can be found in, e.g., Robinson (1994), Smith (2002), and Crane (2011).

[4]Note that metaphysical disjunctivists have different ideas about the category to which illusions belong.

[5]But note that Fish himself is unsure whether this motivation actually works.

also applicable here: if one holds that perception and hallucination are experiences of a different kind, then one might argue on this basis that the evidence in cases of perception and hallucination is also different. That is, one can argue from a metaphysical disjunctivism that upholds a difference in metaphysical kind between perception and hallucination, to an *epistemological disjunctivism* that upholds a difference of evidence in cases of perception and hallucination.

One of the founding fathers of these types of disjunctivism is John McDowell. According to McDowell, there is a good way of responding to arguments from hallucination that attempt to show that perceptions are in an important sense the same as hallucinations:

> But suppose we say—not at all unnaturally—that an appearance that such-and-such is the case can be either a mere appearance or the fact that such-and-such is the case making itself perceptually manifest to someone. As before, the object of experience in the deceptive cases is a mere appearance. But we are not to accept that in the non-deceptive cases too the object of experience is a mere appearance, and hence something that falls short of the fact itself. On the contrary, the appearance that is presented to one in those cases is a matter of the fact itself being disclosed to the experiencer.
>
> (McDowell 1982, p. 80)

McDowell could here be interpreted, and has been interpreted, as proposing a metaphysical disjunctivism according to which the experience itself is different in cases of perception and cases of hallucination. However, especially in later writings (e.g., McDowell 2010, 2013), McDowell has stressed that he only wants to defend the view that the epistemic positions in perception and hallucination are different, and so only wants to uphold an epistemological disjunctivism. This is compatible with the thesis that the experiences in cases of perception and hallucination are exactly the same.

For our purposes, we should also focus on epistemological rather than metaphysical disjunctivism. Even if metaphysical disjunctivism is correct, then epistemological disjunctivism does not immediately follow. Even if one's experiences in cases of perception and hallucination are not of the same mental kind, then this does not yet mean that one's *evidence* in those cases is also different. To get to that conclusion one would first have to assume that one's evidence is partly constituted by one's experiences, an assumption that will already be challenged in the chapters on experientialism. What's more, I hope to show that one can have a perfectly good account of perceptual justification without assuming metaphysical disjunctivism. If this is correct, then this could also be taken as a challenge for metaphysical disjunctivism's purported motivation of preventing skepticism from arising. One need not go to metaphysical disjunctivism to answer skepticism, and even if one does go to metaphysical disjunctivism, then all skeptical worries will not be immediately solved.

The epistemological disjunctivism that I will focus on upholds more than just the thesis that one's evidence is different in cases of perception and hallucination. After all, many views have committed to the idea that there can be differences in evidence in cases that are indistinguishable to the subject. For instance, both William Alston (1988) and Juan Comesaña (2010) take evidence to be necessary for justification, but they think (roughly) that what determines the adequacy of the evidence has to do with whether the evidence reliably indicates the proposition to be justified.[6] Given that one need not be in a position to know whether one's evidence actually reliably indicates what one thinks it indicates, on these views one does not always have access to the soundness of one's evidence. Epistemological disjunctivism, at least as I will use the term, distinguishes itself from these views by also incorporating an accessibility requirement on the soundness of one's evidence.

As I understand epistemological disjunctivism it holds that in paradigm cases of perception, that one *sees* that *p* (rather than merely visually experiences that *p*) is one's *reflectively accessible* evidence for the belief that *p*. This evidence is supposed to be factive in the sense that seeing that *p* entails that *p* is true. Different versions of this view, proposed by respectively Duncan Pritchard (2012b) and Alan Millar (2010), have different conceptions about the precise epistemic role of this evidence and the precise manner of access, but they do agree that this type of factive evidence is present in ordinary cases where we perceive that something is the case. This also accords very well with the datum we started off with, i.e., that we ordinarily appeal to perception in justifying our claims. When our claims are challenged, we do not usually justify them by mentioning that we had a specific type of perceptual experience, instead, we justify them by appealing to a specific perceptual source. We say that we know that person X is at the party because we *see* that person X is standing over there. This gives epistemological disjunctivism a slight advantage in accommodating our ordinary conception of perception and justification in comparison with orthodox experientialist views.

However, epistemological disjunctivism appears to be at a disadvantage with its claim that it is accessible to one that one, e.g., sees that *p*. Doesn't this conflict with the idea that one cannot distinguish a case of hallucination from a case of perception? Pritchard dubs this argument from indistinguishability *The Highest Common Factor Argument*, and portrays it as follows:

P1 In the 'bad' [e.g., hallucinatory] case, the supporting reasons for one's perceptual beliefs can only consist of the way the world appears to one. (Premise)

P2 The 'good' [the cases of veridical perception] and 'bad' cases are phenomenologically indistinguishable. (Premise)

C1 So, the supporting reasons for one's perceptual beliefs in the 'good' case can be no better than in the 'bad' case. (From (P2))

C2 So, the supporting reasons for one's perceptual beliefs can only consist of the way the world appears to one. (From (P1), (C1))

(Pritchard 2008, p. 294)

As Pritchard notes, "[t]he joker in the pack is [...] the move from the second premise to the penultimate conclusion" (ibid.). This move relies on a principle about 'phenomenological indistinguishability', namely that subjects have the same evidence in cases that are phenomenologically indistinguishable. I'll interpret this phenomenological indistinguishability as introspective indistinguishability, which

[6]Note that I will discuss Comesaña's *evidentialist reliabilism* in more detail at the end of Chap. 2.

means that "it concerns an inability on the part of the agent to tell (= know) by introspection alone that case α is non-identical to case β" (Pritchard 2012b, p. 53).[7]

Interestingly enough, Pritchard wants to maintain that good and bad cases are phenomenologically indistinguishable even though one nevertheless has an accessible factive reason available in the good case that is unavailable in the bad case (i.e., that one sees that *p*). Whether these two claims are compatible will depend on what it takes to know by introspection alone that two cases are distinct and on what it means to access a factive reason. Here I just want to remark that we should not assume too easily that hallucination and perception are phenomenologically indistinguishable. What's true is that, when hallucinating, we are usually unable to know on the basis of introspection that the experience we are having is not a perception. However, from this we need not conclude that, when perceiving, we are unable to know on the basis of introspection that the experience we are having is not a hallucination. Although I will reconsider how we would be able to justifiably believe that we are perceiving rather than, say, imagining that *p* on the basis of introspection in Chap. 6, I will not use this as an important argument against epistemological disjunctivism.

What *will* be an important argument against epistemological disjunctivism has to do with our way of accessing the factive reason that we *see* that *p*. It seems that, if this reason is to be accessible, it should be possible for us to entertain higher-order beliefs about ourselves. If I am to use the fact that I see that *p* as a reason to justify my belief that *p*, then I should have the capacity to form the higher-order belief that I see that *p*. Only if I have the possibility of grasping the fact that I see that *p* am I in the position to use this fact as evidence to justify my belief that *p*. But this leads to the problem of *hyper-intellectualization* (Burge 2010): we commonly take unsophisticated perceivers, such as small children and animals, to have perceptual knowledge and hence perceptual justification, but these unsophisticated perceivers do not (yet) have the capacity for higher-order beliefs. If epistemological disjunctivism is correct, unsophisticated perceivers thus turn out, contrary to our ordinary ideas, not to have perceptual justification. This is one of the main arguments against the two varieties of epistemological disjunctivism presented in Chap. 4, although we will see that there are several other worries connected to epistemological disjunctivism as well.

1.4 Process Reliabilism

A good antidote to the problem of hyper-intellectualization has been provided by *externalists* about justification. In contrast to internalists, externalists about justification *deny* that all factors relevant to the justification of a subject's belief are either accessible or internal to that subject. Instead, externalists typically maintain

[7] This definition goes back to ideas presented in Williamson (1990).

that justification is crucially related to factors that are external to the subject, such as the reliability of the belief-forming process (Goldman 1979; Lyons 2009).[8] It is important to notice that externalism about justification is weaker than internalism about justification: even if externalists take, say, reliability to be necessary for justification, they need not deny that accessible evidence also matters in some way. However, if accessible evidence is not necessary for justification, then it is easy to see how externalists can avoid the problem of hyper-intellectualization with regard to perceptual belief. What matters most on such an account is that one's perceptual capacities are properly sensitive to the world, something that can be the case even for very unsophisticated perceivers. There would then be no need for higher-order beliefs about one's own epistemic position.

Process reliabilism is an instance of such a theory, and I will defend a variety of it as providing the most promising account of perceptual justification. According to process reliabilists, perceptual beliefs are justified if and only if they are the undefeated output of a reliable belief-forming process. Given that the perceptual systems of unsophisticated perceivers also reliably produce beliefs about the world, even unsophisticated perceivers can have justified perceptual beliefs. What's more, process reliabilism also allows sophisticated perceivers to easily acquire justification on the basis of perception, although the notion of *defeat* will become more important here. For instance, if I have strong evidence that I have just ingested a hallucinatory drug, or just strong evidence that what I seem to be perceiving cannot be really true, then the ultima facie justification of my perceptual beliefs could be defeated—even if the reliability of the perceptual process is itself sufficient for prima facie justification. However, absent any counter-evidence against the veridicality of my perception, my perceptual beliefs are easily justified. I will have much more to say on the importance of defeat in reliabilist theories of justification in Chap. 6.

So, unlike epistemological disjunctivists, process reliabilists do not encounter a problem of hyper-intellectualization. Indeed, accounting for the possibility of animal knowledge and animal justification has been one of the traditional motivations for externalism (e.g., Burge 2003, 2010). But, given that process reliabilism focuses on reliable perceptual processes rather than higher-order beliefs about the reliability of our perceptual processes, how does it fare in accounting for our ordinary way of talking about perception and justification? Specifically, one might worry that the reliabilist account of perceptual justification fails to explain what is distinctive about the processes of justification that are available to sophisticated perceivers like ourselves.

In fact, this worry can be dealt with. For instance, one way in which reliabilism can account for the fact that we ordinarily cite the deliverances of specific senses to justify our claims is by appealing to our implicit recognition of our senses as reliable sources. When I justify my claim that person X is present by saying that

[8] Note that there are also externalists about *knowledge* who claim that the notion of justification should be construed along internalist lines, but who deny that this type of justification is a necessary condition for knowledge (Kornblith 2008).

I see person X standing over there, I am mentioning a process that is typically accepted as trustworthy because it is reliable. Of course, to be able to provide this justification I do need to be capable of higher-order thought: before I can claim that I see person X standing over there, I need to recognize that I see person X standing over there. But the fact that my providing this sort of justification requires me to have higher-order capabilities does not show that these capabilities are also required to justify my belief in the first place. Instead, process reliabilists will maintain that my perceptual beliefs *are justified* by the reliability of the perceptual process itself, and that my *giving justifications* for these beliefs is a different matter. And one way in which given justifications can be successful is when they cite reliable sources for one's beliefs, assuming that reliability is indeed sufficient for justification. So the reliabilist does have a story which fits our ordinary way of talking about perception and justification.

This leaves one further question for process reliabilism. We saw above that experientialist views were faced with a challenge by the possibility of indistinguishable hallucinations. The challenge did not so much consist in excluding the possibility that one was currently hallucinating on the basis of one's perceptual experience, but rather in explaining why perceptual experience could provide evidence for perceptual (rather than hallucinatory) beliefs in the first place. Epistemological disjunctivism responds by claiming that one's accessible evidence can simply be different in cases of perception and hallucination despite the fact that the cases are introspectively indistinguishable. But what does process reliabilism have to say about introspectively indistinguishable cases of perception and hallucination?

One thing that should be clear is that process reliabilists will also hold that one's epistemic standing can be different in cases that are introspectively indistinguishable. After all, indistinguishability has to do with how things appear from a subject's own perspective, while process reliabilism holds that justification (also) has to do with factors external to a subject's perspective. But even if reliabilism holds that one's epistemic standing can be different in indistinguishable cases, this does not entail that it holds that one's *evidence* is different in these cases. For instance, one could be an eliminativist about evidence, and claim that the only epistemically interesting factors have to do with reliability (Kornblith 2015).

Nor does process reliabilism entail that one's epistemic standing *is* different in *actual* cases of perception and hallucination. A lot will depend on the details of the actual nature of perception and hallucination, and the details of the precise process reliabilist account. For instance, if perception and hallucination constitute very different psychological processes, then perception will presumably come out as a reliable process, while hallucination will come out as an unreliable process. In that case one's perceptual beliefs are justified in actual cases of perception, while one's perceptual beliefs are unjustified in actual cases of hallucination. But we need not dwell too much on that here, as the important point for reliabilists is that indistinguishability does not provide a good guide to either evidence or justification.

1.5 Conclusion

We started this chapter with our ordinary take on perception as a fundamental source of justification. Experientialists take experience to be the crucial source of this justification, where experience is taken as foundational evidence for one's basic perceptual beliefs. This appears problematic as one has to explain how experience is able to fulfill this role given that these experiences could just as well be hallucinations or illusions.

Epistemological disjunctivists hold instead that one's evidence can be different in indistinguishable cases of perception and hallucination, even though this evidence still remains accessible. The type of evidence these epistemological disjunctivists appeal to is constituted by one's seeing that such-and-so is the case. However, one of the major difficulties then lies in spelling out how one is supposed to have access to this fact without creating a problem of hyper-intellectualization.

Finally, process reliabilists appeal to the reliability of a specific sort of process to explain how our perceptual beliefs are justified. Given that the reliability of a process could be inaccessible to its subject, this means that indistinguishability is not a good guide to similarities in evidence or justification.

All of these views will be developed and discussed further in the upcoming chapters. The first two Chaps. 2 and 3, will deal with those accounts of perceptual justification that appeal to *conscious experience* as a necessary and/or sufficient condition for perceptual justification. Chapter 2 focuses on variants of evidentialism, while Chap. 3 focuses on variants of dogmatism. Chapter 4 deals with the epistemological disjunctivist account of perceptual justification, which remains evidential in nature, but changes the evidence from experience to a specific sort of reasons that appear closely connected to *higher-order beliefs*. Chapters 5 and 6 will be devoted to the externalist account of perceptual justification I aim to defend, and focuses in particular on a *non-evidential* process reliabilist account of perceptual justification. Chapter 5 looks at (variants of) the classical account, and Chap. 6 improves on this account by arguing for specific ways of dealing with well-known counterexamples. This improved externalist account of perceptual justification is supposed to incorporate many of the virtues of the other accounts, while still steering clear of their vices.

Chapter 2
Evidentialism and the Problem of Fit

2.1 Introduction

In this chapter we look at the prospects for the first in a family of views that will be labeled as *experientialism*. According to these experientialist views, perceptual justification has to do with having sufficient experiential evidence for one's beliefs. To put it in terms that will be introduced in this chapter: experientialism holds that experience *evidentially justifies* perceptual beliefs.

Before going to a specific way of spelling out this general view, I first discuss what makes it so appealing (Sect. 2.2). This has to do with its ability to explain two important intuitions about perceptual justification, namely, first, the New Evil Demon Intuition, and second, the Blindsight Intuition. The former intuition can be explained by appealing to the sufficiency of perceptual experience for perceptual justification, while the latter can be explained by appealing to the necessity of perceptual experience for perceptual justification. Taken together, these intuitions seem to provide a strong motivation for experientialism.

In Sect. 2.3, I present a well-known problem for experientialism: the Sellarsian dilemma. By using a distinction between evidential and non-evidential justification, this dilemma can be construed as a real problem for experientialism (Lyons 2009). Either experience is non-propositional in nature, and therefore unable to evidentially justify belief at all, or it is propositional, but then it is ad hoc that it can evidentially justify beliefs without being justified itself. This way of setting up the dilemma differs from how it is set up by e.g. Laurence BonJour (1985), and I explicitly discuss this difference.

Section 2.4 focuses on the specific way in which *evidentialism* spells out the experientialist theory of perceptual justification. Evidentialism holds that a certain belief is justified if and only if it fits the available evidence (Feldman and Conee 1985). Applied to perception, this plausibly claims that at least some beliefs are justified if and only if they fit the relevant perceptual experiences. I argue that proponents of evidentialism have failed to provide an adequate notion of fit, given

H. Ghijsen, *The Puzzle of Perceptual Justification*, Synthese Library 377,
DOI 10.1007/978-3-319-30500-4_2

that they want to allow experiences to be non-propositional. In effect, I will be arguing that evidentialism falls prey to the first horn of the Sellarsian dilemma: without propositional content, no sense can be made of the idea that certain beliefs fit certain experiences.

In Sect. 2.5, I go on to discuss an externalist construal of evidentialism (Comesaña 2010). Although this version has a good account of fit, it also has undesirable consequences in that it allows subjects to be justified in cases where they are intuitively unjustified. Moreover, because of its liberal notion of evidence, externalist evidentialism loses the motivation stemming from the Blindsight Intuition.

Section 2.6 summarizes the results of this chapter, and briefly presents how this is related to what will be discussed in the next.

2.2 Two Motivations for Experientialism

Experientialism is the view that perceptual justification is crucially related to having sufficient experiential evidence. In this section I discuss two important motivations for this view, which stem from its ability to accommodate the New Evil Demon Intuition and the Blindsight Intuition. The first intuition can be explained by appealing to the *sufficiency* of experiential evidence for perceptual justification, while the latter can be explained by appealing to the *necessity* of experiential evidence for perceptual justification.

2.2.1 The New Evil Demon Intuition

A first motivation for experientialism comes from reflection on a problem for a simple version of process reliabilism, a view which holds that a belief is justified if and only if it is the output of a reliable process, i.e., a process which tends to produce true beliefs. The problem is known as the New Evil Demon Problem, and was first presented by Keith Lehrer and Stewart Cohen:

> Imagine that, unknown to us, our cognitive processes, those involved in perception, memory and inference, are rendered unreliable by the actions of a powerful demon or malevolent scientist. It would follow on reliabilist views that under such conditions the beliefs generated by those processes would not be justified. This result is unacceptable. The truth of the demon hypothesis also entails that our experiences and our reasonings are just what they would be if our cognitive processes were reliable, and, therefore, that we would be just as well justified in believing what we do if the demon hypothesis were true as if it were false.
>
> (Lehrer and Cohen 1983, p. 192)

The scenario Lehrer and Cohen have in mind is one in which an evil demon or malevolent scientist induces experiences in a subject that are not at all reliably connected to the subject's environment (although, of course, care is taken to prevent the subject from noticing anything about this). For a subject in this New Evil Demon scenario, call him Ned, everything would seem the same as it does to us. He would form beliefs in just the way we would form them, simply by taking perceptual experiences of the world at face value. Yet (simple) reliabilism has the consequence that Ned is unjustified in his beliefs—as they are the output of an unreliable process—while we would be justified in ours. This seems to conflict with our intuition: Ned appears to be as justified as we are. After all, he is forming his beliefs in exactly the same way as we are, and things seem to him exactly the same as they do to us. Call this intuition the *New Evil Demon Intuition*.

Now, Lehrer and Cohen claim that justification "is an evaluation of how well one has pursued one's epistemic goals" (1983, p. 193). They explain the New Evil Demon Intuition by arguing that Ned can be justified in his higher-order belief that his senses are reliable, and consequently in accepting his first-order perceptual beliefs, even if his senses are not, in fact, reliable. But one need not go as far as that, as this explanation immediately brings with it a worry about hyper-intellectualization. Even if subjects (under the influence of a powerful demon) never form higher-order beliefs about their senses, or are incapable of forming those beliefs as in the case of cognitively unsophisticated agents like animals and small children, then we still have the intuition that their first-order perceptual beliefs about their environment could be justified.[1]

A better, alternative explanation of the New Evil Demon Intuition claims that (a) perceptual justification solely has to do with correctly basing one's beliefs on experiential evidence, and (b) that Ned has the same experiential evidence as we do. Why is Ned, like us, justified in believing that there is a table before him? Because, like us, he has a perceptual experience of there being a table before him. The fact that this experience was in his case unknowingly caused by an evil demon does not affect his belief's being justified. This explanation nicely fits in an experientialist's account of perceptual justification, according to which perceptual justification has to do with having sufficient experiential evidence for one's beliefs, so the New Evil Demon Intuition can provide a good first motivation for experientialism.

2.2.2 *The Blindsight Intuition*

A second motivation for experientialism comes from reflection on the phenomenon of blindsight. Some patients with a damaged visual cortex claim not to perceive anything in a certain part of their visual field, but are still able to perform far better

[1]Note that Lehrer and Cohen respond to a hyper-intellectualization objection to their view by arguing that cognitively unsophisticated cognizers may have *information* even though they lack *knowledge* (Lehrer and Cohen 1983, p. 200). However, given that we commonly attribute knowledge to these creatures, an account that could accommodate such knowledge is to be preferred.

than chance when they are forced to guess about, e.g., the presence or orientation of a stimulus.[2] Now suppose that a hypothetical blindsighted subject, call him Bill, is aware of the fact that he hast lost visual experience in a certain part of his visual field, yet—in contrast with actual blindsighters—continues to form correct beliefs about what is going on in his blind field. Bill is not aware of the reliability of his own blindsight capabilities and is unable to produce any more justification for his beliefs other than that it just seems to him that such-and-so is going on in his blind field.[3] Are Bill's beliefs justified? Intuitively, the answer seems to be 'no'. Call this the *Blindsight Intuition*.

Experientialism can easily accommodate the Blindsight Intuition. Bill does not have perceptual justification for his beliefs, because he lacks experiential evidence for them. If he would have visual experiences of what goes on in his blind field, then he would have sufficient evidence for his beliefs. Indeed, this is just what is going on in normal perceivers. However, in the absence of these experiences, Bill lacks sufficient evidence for his beliefs, making them unjustified.

The hypothetical blindsight case resembles one of BonJour's (1985) clairvoyance cases that is meant as a counterexample to externalist theories such as reliabilism:

> Norman, under certain conditions which usually obtain, is a completely reliable clairvoyant with respect to certain kinds of subject matter. He possesses no evidence or reasons of any kind for or against the general possibility of such a cognitive power or for or against the thesis that he possesses it. One day Norman comes to believe that the President is in New York City, though he has no evidence either for or against this belief. In fact the belief is true and results from his clairvoyant power under circumstances in which it is completely reliable.
>
> (BonJour 1985, p. 41)

Again the intuition here is that, contra reliabilism, Norman's belief is *un*justified. Yet the case is still in an important respect different from the blindsight case. In the blindsight case it's clear that if Bill still had intact visual experience, then he would have justification for his beliefs. Something similar is not clearly true for Norman. Even if Norman had some sort of experience that mediated his clairvoyant power, then we might still hesitate to grant that his beliefs were justified. Just suppose that you yourself now suddenly have an extremely vivid image of the President's walking around in New York City. This should not lead you to form the belief that the President is now in New York City. At the very least, it is not as plausible to grant that this belief is justified as it is to grant that Bill's belief is justified if he still had intact visual experiences. This makes the blindsight case better suited as an argument in favor of experientialism, instead of being just an argument against

---

[2]For an overview of the research on, and objections to, blindsight, see Cowey (2010).

[3]This makes Bill importantly different from the super-blindsighters introduced by Daniel Dennett (1991).

reliabilism. The blindsight case more clearly motivates the necessity of experience for perceptual justification, rather than, as the clairvoyant case does, presenting just the insufficiency of reliability for perceptual justification.[4]

2.3 The Sellarsian Dilemma

In the previous section we saw that experientialism could accommodate both the New Evil Demon Intuition and the Blindsight Intuition, thereby deriving a strong motivation in its favor. However, experientialism also faces a general worry about its underlying epistemological theory. Experientialism holds that perceptual experiences provide justification for beliefs by serving as their evidence, even though these experiences themselves cannot and need not be justified. The important question is *how* experiences are able to play this foundational and justifying role. A dilemma, sketched by BonJour (inspired by Sellars 1956), seems to be lurking in the back:

> [...] the proponent of the given [in our case, the experientialist] is caught in a fundamental and inescapable dilemma: if his intuitions or direct awarenesses or immediate apprehensions are construed as cognitive, at least quasi-judgmental (as seems clearly the more natural interpretation), then they will be both capable of providing justification for other cognitive states and in need of it themselves; but if they are construed as noncognitive, nonjudgmental, then while they will not themselves need justification, they will also be incapable of giving it. In either case, such states will be incapable of serving as an adequate foundation for knowledge.
>
> (BonJour 1985, p. 69)

BonJour's thought is that only cognitive states, states with assertive propositional content, are capable of providing justification for belief, and that these states are only able to do so if they are themselves justified. The idea behind this thought is that justification always has to do with having *reasons*, and reasons need to have assertive propositional content to connect them to the beliefs they justify (BonJour 1985, p. 68). However, for any representational state, a question can be raised about its accurateness, which is why they can only confer justification if they are themselves justified (BonJour 1985, p. 78). Something can act as a reason only if one is already justified in taking the content of this reason to be true.

As Jack Lyons (2009, p. 40) points out, there are some difficulties with BonJour's argument. It's not clear why justification must always involve having reasons, nor is it clear why representational states must be justified before they can confer justification. Even if reliability is not necessary or sufficient for justification, we do have the intuition that the reliability of a certain process can be relevant for the

[4]Something similar can be said in relation to Lehrer's (1990) Truetemp example, in which Mr. Truetemp unknowingly gets a so-called tempucomp implanted in his brain which then reliably produces beliefs about the specific temperature. Even if the tempucomp works by producing certain experiences in Truetemp, then it is still unclear whether his beliefs are justified.

justification of the belief that is its outcome. Similarly, even if the coherence of a set of beliefs is not necessary or sufficient for these beliefs to be justified, the coherence might well be taken to improve the individual justification of each of the set's members. Yet reliability or coherence by itself is not a reason for or against a certain belief, nor is it the kind of thing that can be justified. So why assume that *all* justification has to do with having reasons, and that *all* justification presupposes that the justifier is itself justified?

To strengthen the Sellarsian dilemma for experientialism, Lyons distinguishes between evidential and non-evidential justifiers (2009, pp. 21–24). An evidential justifier for a belief "is any state that serves as part or all of the agent's justifying grounds, that is, evidence, that is, reasons for that belief" (Lyons 2009, p. 23), while a non-evidential justifier is "anything else that makes it the case that the belief is justified, or contributes toward making it the case that the belief is justified" (Lyons 2009, p. 24). This will enable us to count reliability and coherence as non-evidential justifiers for beliefs, while, for instance, beliefs could count as evidential justifiers for other beliefs. Whether these non-evidential and evidential justifiers in the end succeed in justifying a belief depends on other factors as well. Beliefs held for bad reasons will not justify other beliefs, and coherence alone will not provide justification if the set of coherent beliefs is totally disconnected from the external world.

Evidential justifiers need not be taken as such by agents basing their beliefs on them. If a certain belief is used as a basis for another, I need not be aware of this relation. For instance, having performed the same inferential steps several times, I might come to a justified conclusion without consciously reasoning through all of the needed premises. In such a case, it seems that I was not aware of all of the evidential justifiers that were relevant in justifying my conclusion. However, in this case it also seems that I could become aware of those evidential justifiers if I reflected on how I arrived at the conclusion. But even this need not be the case. Suppose that I have unconscious biased beliefs against certain groups of people. Then I might actually base my decision against hiring someone from this group on this unconscious belief, all the while thinking that I based it on something else instead. What matters for being an evidential justifier is that a state is actually *used* as evidence for another belief, rather than that an agent is aware of so using this state.

On the basis of the distinction between evidential and non-evidential justifiers we can define evidential justification as the relation between an evidential justifier and the beliefs it justifies, and non-evidential justification as the relation between a non-evidential justifier and the beliefs it justifies. However, according to Lyons, it is important to realize that there is only one property of being justified, although there are two different relations things might have towards a belief with this property (Lyons 2009, pp. 25–6). Non-evidential justifiers are supposed to be the *metaphysical basis* for justification, and are thus present in any case of justification. Even if a belief that p is properly based on the belief that q, then the fact that the belief is so based is still supposed to count as a non-evidential justifier. So if one were to think of the property of being justified as coming into two separate kinds, the non-evidential and evidential kind, then all beliefs would come out as being non-evidentially justified, and only some beliefs would in addition be evidentially justified.

Still, one might as well hold that the fact that a belief is properly based on another belief is different in kind from the fact that a belief was produced by a reliable cognitive process. Perhaps one would then prefer to say that being non-evidentially justified *is* different in kind from being evidentially justified. This would go more into the direction of, for instance, Tyler Burge (2003) and Peter Graham (2012), who both claim that the genus of *epistemic warrant* has *entitlement* and *justification* as its species, where only justification has to do with having reasons or evidence for one's beliefs. I'm also fine with taking the distinction between evidential and non-evidential justification along these lines. Whichever way one wants to draw the distinction, the important point is that there is a distinction to be made: some factors simply impact the justification of a belief without serving as reasons or evidence for that belief.

The earlier observations with regard to BonJour's argument can now be expressed in these terms. We need not assume that all justification has to do with reasons, as there might be beliefs that are only non-evidentially justified (e.g., by being the output of a reliable process). Similarly, not all justification will presuppose that the justifier is itself justified, as non-evidential justifiers like reliability and coherence are not even the kinds of things that can be justified. But we can reformulate BonJour's argument in terms of evidential and non-evidential justification, making it more plausible.[5] The dilemma for the experientialist then becomes the following: either experience is non-propositional, but then it cannot *evidentially* justify belief at all; or experience is propositional, but then it would be ad hoc to claim that it could evidentially justify beliefs without being justified itself. Note that the claim is not that experience needs to be *evidentially* justified before it can evidentially justify belief. The idea is precisely that non-evidential justifiers provide the ultimate foundation for evidential justifiers.

This version of the Sellarsian dilemma has more bite to it, as it specifically attacks the experientialist's claim that experiences act as *evidence* for belief. First, it makes sense to suppose that all evidence is propositional, because evidence needs to stand in some sort of logical relation to the propositions it supports. How else could we explain why precisely those propositions are supported by precisely that evidence? This is what lends credibility to the first horn of the dilemma: if experience is non-propositional, then it is entirely unclear how it would be evidence for some beliefs but not others.

Second, it makes sense to suppose that propositional states need to be justified before they can be rightly considered as evidence, or at least, evidence capable of justifying beliefs. Not just any propositional state can count as evidence for another. For instance, if I simply adopt some beliefs that logically support another, then

[5] Note that the way I present the Sellarsian dilemma is different from the precise way in which Lyons construes the dilemma himself.

these beliefs still do not count as evidence for the belief they support. Further, desires and imaginations also do not count as evidence even though they might both be propositional states. Both facts can be explained by appealing to the idea that propositional states need to be justified before they can be evidence for another. This lends credibility to the second horn of the dilemma: if experience is propositional, then it is ad hoc to claim that it can evidentially justify beliefs without being itself justified.

In the next section, I will expound on the first horn of this version of the Sellarsian dilemma by discussing evidentialism, a theory which holds that experiences can act as evidence for beliefs even if they are non-propositional. The second horn of the dilemma will await further discussion until the next chapter, when the dogmatist theory of perceptual justification is at issue.

2.4 Evidentialism

One well-known theory which can be understood as grasping the first horn of the Sellarsian dilemma is Richard Feldman's and Earl Conee's evidentialism. Feldman and Conee characterize their evidentialist position about justification with the help of the following biconditional:

Evidentialism Doxastic attitude *D* toward proposition *p* is epistemically justified for *S* at *t* if and only if having *D* toward *p* fits the evidence *S* has at *t*.

(Feldman and Conee 1985, p. 15)

The first thing to notice about this characterization is that it is only about what is called *'propositional justification'*: it tells us only which doxastic attitude is justified for a subject, and not whether a subject is actually justified in the doxastic attitude he has. The latter notion is called *'doxastic justification'*. A certain doxastic attitude might be propositionally justified without being doxastically justified if the subject in question does not base the attitude on the evidence that makes it propositionally justified. To give an example: if I have strong evidence for the belief that *p*, but do not believe *p* on the basis of that evidence (I rather like the sound of *p* and therefore believe it), then my belief that *p* is propositionally but not doxastically justified.

To accommodate the notion of doxastic justification, Feldman and Conee also present a characterization of well-foundedness (WF):

WF *S*'s doxastic attitude *D* at *t* toward proposition *p* is well-founded if and only if

(i) having *D* toward *p* is justified for *S* at *t*; and
(ii) *S* has *D* toward *p* on the basis of some body of evidence *e*, such that
    (a) *S* has *e* as evidence at *t*;
    (b) having *D* toward *p* fits *e*; and
    (c) there is no more inclusive body of evidence e' had by *S* at *t* such that having *D* toward *p* does not fit e'.

(Feldman and Conee 1985, p. 24).

This characterization includes that a subject actually has a certain doxastic attitude on the basis of some body of evidence, and thus enables evidentialists to deal with the above distinction between propositional and doxastic justification. There are still some important questions left unanswered though. First, what *constitutes* the body of evidence for a doxastic attitude, and, second, when does a doxastic attitude *fit* the evidence?

With regard to the first question, Feldman and Conee answer that "it seems clear that this [evidence] includes both beliefs and sensory states such as feeling very warm and having the visual experience of seeing blue" (Feldman and Conee 1985, p. 32, n. 2). However, it seems equally clear that not *all* beliefs should count as evidence. I cannot improve the epistemic status of my belief by randomly adopting other beliefs that support it. One plausible explanation of this fact is that the randomly adopted beliefs would not be justified, and therefore unable to confer any justification onto the beliefs that can be inferred from them. So an important constraint on those beliefs that can constitute evidence seems to be that these beliefs must themselves be justified, i.e., supported by evidence. Although this looks like the start of a regress, note that this need not be the case if sensory states can also constitute evidence. This would make sensory states extremely important for Feldman and Conee's theory, as they would provide the foundation for all other evidence.[6]

However, sensory states also bring with them some difficult issues if they are to serve as evidence. Feldman and Conee do not require that sensory states must have propositional content to constitute evidence, as they do not really think that experiences take propositions as their objects. According to them, such a view is one of several "somewhat contorted attempts to make some other views match the view we prefer" (Conee and Feldman 2004a, p. 2). But if sensory states do not have propositional content, then it is unclear how they, on their own, could serve as evidence for specific propositions. Indeed, it is precisely for this reason that John McDowell (1994) maintains that experiences should have propositional content: only then can experiences constitute genuine reasons for beliefs.

Feldman and Conee are surprisingly silent about this matter. What is it about a feeling of warmth that makes it evidence for the belief that I am warm, instead of evidence for the belief that I am cold? If the state does not have propositional content, then it certainly does not imply that I am warm instead of being cold. Nor does it seem to make sense that the experience itself makes a certain belief probable.[7] The proposition *that* an experience with such-and-such properties occurs

[6]Couldn't Feldman and Conee also adopt coherentism instead? They could, but the problems for coherentism are well-known: coherent systems need not be true—it seems that they could lack any connection to the world whatsoever—and there is again the danger of over-intellectualizing how perceptual beliefs are justified. See Pollock (1974, pp. 26–29) for an early presentation of some of these objections.

[7]See Williamson (2000, pp. 194–200) for this kind of argument in favor of the thought that all evidence is propositional.

might stand in these relationships to belief, but this proposition is different from the experience itself.

James Pryor (2005, p. 193) thinks that there might be another way to find a connection between non-propositional experience and belief that explains how the former could justify the latter. This suggestion has to do with the "logical structure" of events and propositions, and the idea is that the event of my having an experience, e.g., a headache, has a similar logical structure as the proposition that I have a headache. This similarity in logical structure supposedly explains why my having a headache justifies my belief that I have a headache, instead of some other proposition. Unfortunately, at first glance it seems that the event of my having a headache is again not the same as the experience of the headache itself. The event of my having a headache seems to include me *and* the headache that I'm having, whereas the experience of the headache might not include me in the same way. If this is true, then we would no longer retain Feldman and Conee's idea that experiences themselves can act as evidence for belief.

A more important problem has to do with the fact that experiences are not only supposed to provide evidence for introspective beliefs, but also for external world beliefs. Should we then say that the logical structure of, e.g., the event of my having a visual experience of a laptop in front of me is similar to *both* the logical structure of the proposition that I have a visual experience of a laptop in front of me *and* the logical structure of the proposition that there is a laptop in front of me? What's more, it also seems possible that an experience of a red, roundish shape in front of me justifies my belief that there is a tomato in front of me, but here the similarity in logical structure is even more difficult to find. One could of course maintain that experiences only justify beliefs about, say, shapes and colors, but then the experientialist proposal will certainly lose some of its appeal.

Another suggestion to link non-propositional experiences to beliefs is provided by Laurence BonJour (2003). BonJour takes there to be a descriptive relation between a non-propositional experience and a conceptual belief *about* that experience, where the character of the experience determines whether the description is true. BonJour further claims that "an awareness of that non-conceptual [and non-propositional, p. 70, n. 6] character can seemingly constitute a kind of *reason* for thinking that the description is true or correct [...]—thus apparently providing a basis for the justification of the conceptual claim" (BonJour and Sosa 2003, p. 72). Now if BonJour means that we have a higher-order awareness of our experience, then this just pushes the question about evidence and propositionality one step further back. How could this higher-order awareness constitute evidence for a specific belief if it is non-propositional (or if it is propositional: why is this higher-order experience itself not in need of justification)?

However, BonJour stresses that the awareness should not be thought of as a higher-order awareness of the experience (BonJour and Sosa 2003, pp. 63–4), but rather, as a "constitutive, or "built-in," non-apperceptive awareness of [the experience's] distinctive sort of [sensory] content" (BonJour and Sosa 2003, p. 70). But then this suggestion seems to come down to little more than *stating* that being conscious of a non-propositional content, in other words, undergoing a non-propositional experience, can serve as evidence for belief—albeit a belief about the experience itself. We started with the question of how a non-propositional experience could constitute evidence for a specific belief, and the answer we've been given claims roughly that our first-person awareness of an experience constitutes a reason for holding a specific belief about that experience. This just leaves unexplained how this awareness is supposed to favor one belief over another.

A final internalist suggestion to overcome the problem of providing an evidential connection between non-propositional experience and belief would have it that the feeling of warmth, together with my knowledge of the reliable connection between this feeling and my being warm, is sufficient to constitute evidence of my actually being warm. However, such a move would only invite the question where the knowledge of the reliable connection stems from. On what evidence is this knowledge in turn based? We would now be back in the evidential regress that sensory states were exactly meant to end.

One might therefore turn (slightly) externalist instead, and hold that a feeling of warmth is evidence for the belief that I am warm merely because it reliably indicates that I am, thereby no longer presupposing that I have to know about this connection. This cannot explain how experience provides *doxastic* justification though, or at least, it cannot on Feldman and Conee's definition of it (WF). At best this might provide a way to define propositional justification: if a subject has access to a reliable indicator (evidence) that *p*, then he has propositional justification for the belief that *p*. But if the subject does not know of the connection between the reliable indicator and what it indicates, then he can hardly be said to hold his belief that *p* *"on the basis of"* (Feldman and Conee 1985, p. 24) this reliable indicator—after all, we supposed that the subject need not know of the reliable connection between the indicator and *p*.

Note that the problem here is not the fact that the supposed subject does not consciously take his feeling of warmth to be a reliable indication of the fact that he is warm. As mentioned before, one can base one's belief on some evidential justifier without realizing that one is doing so, as can be the case with unconscious biases. The problem in this case is that, even when the subject is presented with his supposed evidential justifier, i.e., the reliably indicating non-propositional experience, then he would still not be able to understand its connection to what it reliably indicates in the absence of any knowledge of this connection. This might not be obvious in the case of feeling warm, but it becomes more obvious when we look at different examples. For instance, take the following example from Jack Lyons:

> Suppose a mad neurosurgeon rearranges some of my neural connections while I sleep, in such a way that excessive pressure on my right big toe now reliably causes a sensation of warmth on my left cheek, but for some reason, I'm prone to infer from this that there's pressure on my right big toe. Suppose that the cheek sensation only occurs when there is pressure on the toe. Supposing that I have no idea that this odd connection holds, the intuitive verdict is that the belief is not prima facie justified.
>
> (Lyons 2009, p. 64)

If one supposes that experiences can do their evidential work without having propositional content, but simply on the basis of reliable indication, then the example should be one in which the subject does turn out to have a justified belief. Yet this is clearly not the case, because it seems that the evidential connection should somehow be understandable by the subject in question. This is an important difference with the case of unconscious bias: once someone tells me that my belief that the candidate is unfit for the job is based on a biased belief I have against this candidate, then I will be able to understand the evidential connection between the biased belief and the opinion about the candidate—even if I still don't accept that I actually have this biased belief. Being able to understand evidential connections does not mean that one is always aware of them.

The only process that could be going on between the reliable indicator and the belief, in the absence of knowledge about the connection between indicator and what it indicates, seems to be a purely causal rather than evidential process. Of course we could choose to use 'basing' and 'evidence' in this externalist way, but this would then describe a justificatory process that is quite different from standard cases of evidential justification, where, say, someone believes that q on the basis of the belief that p and the belief that if p, then q. So instead of providing an illuminating account of perceptual justification, this way of putting matters would actually obscure important differences between cases of evidential and non-evidential justifiers.

So far we have only considered the question of how a single non-propositional state could constitute evidence for a specific proposition. The situation only gets worse once one focuses on the question of when a doxastic attitude fits an entire body of evidence. The idea is that a complete body of evidence could favor or disfavor (or neither) a certain proposition. But then some sort of logical interaction between different pieces of evidence seems to be necessary to determine whether the complete body of evidence favors a proposition or not. Now suppose that there are non-propositional sensory states among the complete body of evidence. Then it is entirely unclear how these should be weighed in combination with, say, propositional beliefs. If I believe that the store is open on Monday, but you tell me that the store is closed on Monday, then your testimony contradicts my belief. My belief has the content that p, while your testimony has the content that $\neg p$. But there would be no such contradiction in content between my feeling of warmth and my belief that I am cold. If there is a different kind of tension between these two things, then it would be nice to have some sort of theory about it. And once one even allows feelings of confidence or certainty into the evidence mix (which is hinted at several times by Conee and Feldman[8]), then there even is an extra non-propositional factor to make the weighing process more difficult.

The problem of making clear how a non-propositional state could serve as evidence for a specific proposition, and the problem of explaining how a certain doxastic attitude could fit an evidence set which consisted partly out of non-propositional states, appear to be serious problems for the evidentialist view as proposed by Feldman and Conee. Note that my claim is not that non-propositional sensory states could never contribute to the justification of beliefs. My claim is

[8]See their collection of essays (2004a), pp. 70, 76, 112, 238–9.

merely that it is entirely unclear how non-propositional states could *evidentially* justify certain beliefs—exactly the worry raised by the first horn of the above Sellarsian dilemma. This leaves open whether they contribute to the justification of beliefs without serving as evidence for them by, for instance, reliably causing them. Indeed, the theory I will defend in the final chapters is fully compatible with this idea.

2.5 Evidentialist Reliabilism

In the previous section we saw that evidentialism had major problems in explaining how non-propositional states could evidentially justify beliefs. Juan Comesaña has recognized some of these problems, and proposes a specific synthesis of reliabilism and evidentialism, called *evidentialist reliabilism*, to solve them (Comesaña 2010).[9]

First of all, Comesaña proposes the following clear alternative account of fitting the evidence:

Reliabilist Fit Necessarily, believing that p fits e for subject S if and only if:

1. e doesn't include any beliefs of S and the connection between S's having e and p is actually reliable; or
2. e includes beliefs of S, all of these beliefs are justified, and the connection between S's having e and p is conditionally actually reliable.

(Comesaña 2010, p. 581)

Notice that Comesaña distinguishes between a reliable connection and a conditionally reliable connection between S's having e and p's being true.[10] The former requires that S's having e is mostly connected to p's being true, while the latter requires that S's having e is mostly connected to p's being true *given that all of the beliefs in e are true*.

Comesaña gives a similar account of when disbelieving that p fits the evidence (just substitute $\neg p$ for p in the two clauses), and defines suspension of judgment as a fitting attitude when neither believing nor disbelieving fits the evidence. The benefit of this account of evidence-fit is that it need not consider any logical connections between different pieces of evidence, but can instead focus on the connection between the subject's *having* some collection of beliefs and sensory states, and the truth (or falsity) of the proposition in question. Thus, the previous problem of evidence-weighing disappears.

Yet this externalist account of fit also encounters similar worries as the earlier externalist suggestion, having to do with a subject's own perspective on his

[9] It is also meant to solve some problems for reliabilism, to wit, the problem of Clairvoyance (mentioned in Sect. 2.2.2), and the Generality Problem (which will be discussed in Chap. 5).

[10] For our purposes we won't focus on Comesaña's point that the connection needs to be (conditionally) *actually* reliable, where 'actually' is understood as having a two-dimensional semantics (Comesaña 2010, p. 579).

evidence. This will be clearer when we look at the full theory developed by Comesaña:

Evidentialist Reliabilism A belief that p by S is justified if and only if:

1. S has evidence e;
2. the belief that p by S is based on e; and either

   (a) e doesn't include any beliefs and the type *producing a belief that p based on evidence e* is actually reliable; or
   (b) e includes other beliefs of S, all of those beliefs are justified and the type *producing a belief that p based on evidence e* is conditionally actually reliable.

(Comesaña 2010, p. 584)

Now consider the following case. A subject is told by a reliable source that p is the case. In being told that p is the case, the subject presumably undergoes a specific sort of auditory experience. However, the specific auditory experience in this case is, unbeknownst to the subject, actually reliably connected with $\neg p$ because the specific tone of voice indicates that the source is lying. Let's further suppose that the subject has no beliefs about the reliability of the source to make sure that condition (b) does not spring into action (this is a simplifying assumption that could probably be dropped). What results would Comesaña's account have in this specific case?

First, let's look at what Comesaña's account predicts about evidence-fit. For now, let's suppose that the only relevant evidence is the auditory experience in question. Given that having this specific auditory experience is reliably connected to $\neg p$, disbelieving that p would now fit the evidence. There seems to be no problem with talking this way, given that we are viewing things from the externalist perspective.

Let's go on to look at what beliefs will be justified for the subject under consideration. Suppose the subject, not unnaturally, believes that p because of what his source told him. The type in question, believing that p on the basis of having this specific auditory experience, is actually unreliable. So the subject does not have a justified belief that p. Again, from an externalist perspective this might not be such a bad result: although the subject himself thinks he has a justified belief, the external conditions are such that he does not actually have a justified belief. The subject might be blameless in his belief, but not, unfortunately, justified.

However, suppose now that the subject actually believes that $\neg p$ on the basis of his auditory experience. He typically does not do so when people tell him that p, but in this particular case, he just happens to do it. The belief-forming type in question now becomes reliable, because $\neg p$ is actually reliably connected to the specific evidence. This means that, on Comesaña's account, the subject under consideration turns out to have a justified belief that $\neg p$. But even from an externalist perspective, this is an unintuitive result. The subject has no reason to base exactly that belief on the evidence he has, at least when the evidence is viewed from his own perspective.

Note that it won't help Comesaña's account if one adds additional beliefs to the evidence set e. For instance, suppose that the subject not only has the specific auditory experience he has, but also believes that he has just been told that p. We can assume that this belief is justified on the basis of the auditory experience, and so it can rightfully be taken to be part of the evidence. Given that there are now beliefs in *e*, we should take into consideration clause (b) which deals with conditional reliability. However, this new evidence set *e* is also conditionally reliably connected to $\neg p$: even when it is true that one has just been told that *p*, the specific auditory experience in question will still be reliably connected to $\neg p$ rather than *p*. So if the subject believes that $\neg p$ on the basis of his auditory experience and his belief that he has just been told that *p*, then the subject's belief is still justified according to Comesaña's account. Of course, the result one would like to have is that the subject's belief that he has been told that *p* works as a *defeater* for the justification of his belief that $\neg p$. But this result simply cannot be had in Comesaña's theory.

Another response to the example under consideration might focus on the fact that testimonial justification cannot be reduced to justification that arises out of auditory experiences. Where the example has gone wrong, according to this response, is in its supposition that testimonial justification can be reduced to perceptual justification. Although this non-reductivism about testimony might be correct, this does not seriously affect the example against Comesaña's account. First of all, whether or not one wants to classify it as *testimonial* evidence, the auditory experience remains some sort of evidence for $\neg p$. Second, even if one wants to include the perhaps irreducible evidential fact that someone testified that *p* in the evidence set, then one is left with an evidence set that will, in its entirety, still be reliably connected to $\neg p$ rather than *p*. So the problem remains.

Finally, one might also think that all of this is just an instance of the *generality problem*: what exactly is the type of belief-forming process at work here?[11] Is it believing on the basis of testimony, believing on the basis of auditory experiences, believing on the basis of these specific auditory experiences, etc.? However, Comesaña's account is exactly meant as a way of responding to this problem, as the relevant belief-forming types in his account are specified as *producing a belief that p based on evidence e*.

Instead of appealing to the generality problem, I suggest that what has gone wrong has to do with the combination of a very liberal externalist conception of evidence and a rather vague notion of basing. To compare: it's one thing to say that the growth rings of trees are evidence for a tree's age because they reliably indicate that age, but quite another to say that a subject is evidentially justified in believing that the tree is of a certain age *on the basis of* the amount of growth rings if the subject does not have any justified beliefs about the connection between growth rings and age. What one could do to overcome this problem is to restrict the notion of 'evidence' or 'basing' in such a way that a subject is required to have justified beliefs about the relation between the evidence and the beliefs it supports. But of course this merely leads back to the regress we encountered earlier: if perceptual beliefs require justified higher-order beliefs in order to be justified, then it is totally unclear how these higher-order beliefs in turn could be justified.

[11] I will discuss this problem in more detail in Chap. 5.

The fact that the notions of 'evidence' or 'basing' are, without further restriction, problematic for evidentialist reliabilism also shows up once we consider one of its purported motivations.[12] Comesaña believes that his evidentialist reliabilism, unlike alternative reliabilist accounts, is able to escape the Clairvoyance Problem mentioned in Sect. 2.2.2. According to Comesaña, "[...] one crucial feature of BonJour's example is that Norman has no evidence for or against his clairvoyant powers, or regarding the whereabouts of the President—the belief just "pops up" in his head" (Comesaña 2010, p. 582). Evidentialist reliabilism is supposed to solve this problem by explicitly making evidence necessary for justified belief, thereby excluding Norman from having a justified belief regarding the whereabouts of the President. The problem is, though, that with Comesaña's liberal conception of evidence, there is no reason to assume that Norman did not have any evidence for his beliefs. Norman's clairvoyant faculty is assumed to work reliable, so there will probably be some cognitive mechanism at work that tracks, say, the changes in energy waves and outputs beliefs about the future. Nothing excludes that Norman has unconscious representations that reliably indicate that such-and-such is going to happen, so nothing excludes that Norman has evidence that such-and-such is going to happen.[13]

For the same reason, evidentialist reliabilism cannot be motivated by appealing to the Blindsight Intuition mentioned in Sect. 2.2.2. According to the Blindsight Intuition, Bill, our hypothetical blindsighted subject, lacks justification for his perceptual beliefs, and evidentialism could explain this by pointing to the fact that Bill lacks experiential evidence. But if having evidence that *p* merely requires that a subject has a certain mental state which reliably indicates that *p*, then Bill can still be said to have evidence that *p*. This means that evidentialist reliabilism, construed with a liberal notion of evidence, cannot explain why Bill would have unjustified perceptual beliefs.

To overcome this problem, evidentialist reliabilism would have to incorporate some kind of an extra requirement on evidence. A consciousness requirement which says that all evidence has to be conscious, seems too strong: surely an evidentialist would want to allow that one can base beliefs on evidence without consciously inferring those beliefs from the evidence. For instance, a mathematician who immediately 'sees' the solution to a certain difficult mathematical problem should turn out to have based his belief on his previous mathematical knowledge—even if he was not consciously aware of it.

An accessibility demand on evidence seems more promising, but has its own problems. For instance, one could say that a subject has evidence that *p* if and only if he is able to have introspective access to a certain mental state that reliably indicates

[12] I will focus on the notion of evidence in my discussion, but one could equally make the point by asking for the correct conditions for basing a belief on the evidence.

[13] In fact, as remarked above (Sect. 2.2.2), even if Norman were to have some kind of evidence in the form of a specific sort of clairvoyant experience, then he might still appear to have unjustified beliefs.

that *p*. But now the notion of 'access' seems to set us in the direction of the earlier regress again. Having introspective access to a certain mental state seems to be best spelled out in epistemic terms, namely, that one has some justified beliefs about that mental state acquired via introspection (Fumerton 1995, p. 64). But where does the evidence for these justified beliefs then stem from? What's more, one might not want to take a stand on the empirical thesis that all of our perceptual beliefs are based on such accessible evidence: why not think that we can e.g., recognize things and people for what they are without having access to the properties on the basis of which we recognized them?

These considerations seem to be sufficient to place a large burden of proof on proponents of evidentialist reliabilism. Without a good refinement of the notion of 'evidence' (or of 'basing'), evidentialist reliabilism will have counter-intuitive results and will not be motivated by the Blindsight Intuition. Moreover, the necessary refinement does not appear to be forthcoming, given that some of the most likely candidates lead to problems. In combination with the fact that standard evidentialism has no satisfactory account of fit, the non-propositional evidentialist take on perceptual justification is definitely in trouble.

2.6 Conclusion

I have started this chapter with describing two motivations in favor of experientialism, the view that perceptual justification has to do with having sufficient experiential evidence for one's beliefs. These motivations arise from its ability to accommodate two intuitions about perceptual justification, to wit, the New Evil Demon Intuition and the Blindsight Intuition. By appealing to the sufficiency of experiential evidence for perceptual justification, experientialism is able to explain why radically deceived subjects like Ned are still justified in their perceptual beliefs. By appealing to the necessity of experiential evidence for perceptual justification, experientialism is able to explain why a blindsighted subject like Bill lacks perceptual justification.

However, experientialism also faces a version of the Sellarsian dilemma. On the one hand, if experiences are construed as non-propositional, then it is entirely unclear how they can *evidentially* justify perceptual beliefs. On the other hand, if experiences are construed as propositional, then it appears ad hoc to claim that they can evidentially justify beliefs without being in need of justification themselves. Although there surely are instances of justification in which the justifier is not itself propositional or justified (think of reliability or coherence), it is also clear that these are not cases of evidential justification. This makes this version of the Sellarsian dilemma, in terms of evidential and non-evidential justification, much harder to refute.

Evidentialism grasps the first horn of this Sellarsian dilemma. It holds that a subject has a justified belief if and only if it is based on evidence that fits the doxastic attitude of belief. For some cases of perception, this plausibly comes down to the idea that a subject has a justified belief if and only if it is based on non-propositional experience. Unfortunately, evidentialism is unable to overcome the problems associated with the horn of non-propositionality. First, it does not explain how certain non-propositional sensory states could serve as evidence for some beliefs but not others in the absence of propositional content logically linking the states to the beliefs. Second, it does not explain how to determine which doxastic attitudes fit the evidence, given that the evidence can consist partly out of non-propositional states.

Evidentialist reliabilism is meant as an improvement of reliabilism by employing a reliabilist notion of evidence-fit. However, without restricting the notion of evidence far more than its reliabilist roots would suggest, evidentialist reliabilism will give counter-intuitive verdicts about justification and will lose the motivation stemming from the Blindsight Intuition. Moreover, the correct restriction of the notion of evidence is not easily available, as two obvious candidates, consciousness and accessibility, both lead to problems. Thus, evidentialism appears in trouble on either variant of it.

This naturally leads to the question of whether it would be better to go for the second horn of the Sellarsian dilemma. Perhaps a version of experientialism can be upheld if we assume experiences to be propositional. The next chapter will focus on one popular internalist alternative that tries to develop such a view: dogmatism.

Chapter 3
Dogmatism and the Distinctiveness Problem

3.1 Introduction

In the previous chapter I introduced experientialism as the view that perceptual experience evidentially justifies beliefs. Although this view was motivated by the New Evil Demon Intuition and Blindsight Intuition, it faced a dilemma in how it construed perceptual experience. Either experience is non-propositional, but then it is entirely unclear how it could evidentially justify beliefs; or it is propositional, but then it is ad hoc to hold that it can evidentially justify beliefs without being justified itself. Evidentialism succumbed to the first horn of the dilemma by being unable to explain the supposed evidential relation between non-propositional experience and belief. In this chapter I present the perhaps most popular variant of experientialism that grasps the other horn of the dilemma: dogmatism.

According to dogmatism, if one has a perceptual experience that *p*, then one thereby has immediate prima facie justification for the belief that *p* (Pryor 2000; Huemer 2001; Chudnoff 2011).[1] There are three important parts to this variety of dogmatism: a dogmatist part, a phenomenalist part, and an experientialist part. I will explain each of these parts in Sect. 3.2.

In Sect. 3.3, I look at two problems for dogmatism that force it to refine its notion of perceptual experience. The first problem has to do with distinguishing between novice and expert perceptual justification, while the second is known as the problem of the speckled hen. Some dogmatists have responded to these problems

[1]Note that Huemer is concerned with a general theory of justification based on seemings that *p* (called phenomenal conservatism), Pryor with a specific theory of perceptual justification, and Chudnoff with a specific theory of intuitive justification.

H. Ghijsen, *The Puzzle of Perceptual Justification*, Synthese Library 377,
DOI 10.1007/978-3-319-30500-4_3

by distinguishing perceptual seemings from perceptual sensations, restricting their claim about justification to the seemings. I argue that there is no reason to assume that seemings and sensations exist as separate conscious mental states, and that dogmatists are better off with identifying perceptual experiences with conscious states that incorporate aspects from both sensations and seemings.

However, even with this refined notion of perceptual experience in hand, dogmatism still faces the problem outlined in the second horn of the Sellarsian dilemma, which I call the 'Distinctiveness Problem'. Dogmatism needs to explain what is so distinctive about perceptual experience that enables it to evidentially justify beliefs (in contrast with e.g., desire and imagination) without being in need of justification itself (in contrast with belief). I discuss this problem in Sect. 3.4, and look at several suggestions on behalf of dogmatism that could be used to answer it. All of these mention a purported phenomenal property of experience that explains its distinctive justificatory power. Unfortunately, these properties are either (a) not specific to perceptual experience, or (b) reducible to, or possibly caused by, higher-order beliefs. Moreover, it's unclear why phenomenal features would have any epistemic significance at all.

In Sect. 3.5, I elaborate on this general critique of dogmatism by presenting an alternative version of the speckled hen problem. Although refining the notion of perceptual experience solves the first version of this problem, it does not help to solve this alternative version. It shows that having a perceptual experience that *p* will not be sufficient to justify the belief that *p* if reliability is not somehow incorporated as an extra requirement of the theory. But once one does incorporate reliability into the theory, the New Evil Demon Intuition no longer motivates dogmatism, and it furthermore becomes more plausible that reliability is already sufficient for justification by itself.

In Sect. 3.6, I briefly respond to an argument raised by Michael Huemer (2007) which attempts to show that dogmatism is the only viable account of perceptual justification because we simply base all our beliefs on the way things seem. I respond by noting that the argument suffers from a fallacy of ambiguity and that its premises need two different notions of basing to be plausible.

The final Sect. 3.7 sums up the results of this chapter and briefly looks ahead to what will be discussed in the next.

3.2 Phenomenalist Dogmatist Experientialism

Dogmatism about perception is usually presented along the following lines:

> **Dogmatism** Having a perceptual experience that *p* is sufficient for immediate prima facie justification of the belief that *p* (Pryor 2000; Huemer 2001).

Sometimes it is expressed in slightly different terms as, for example, the view that "if it perceptually seems to you that *p*, then you thereby possess some prima facie justification for believing that *p*" (Chudnoff 2011, p. 314).[2] Both the *immediacy* and *defeasibility* of justification are important aspects of the view.

Starting with the former notion, according to dogmatism, perceptual justification is immediate in the sense that it "doesn't rest on any other evidence or justification you have for believing other propositions" (Pryor 2000, p. 532). The thought is that perceptual experience can provide justification for the belief that you have two hands even if you do not have any prior justification for the belief that you are not a brain in a vat. Dogmatism is thus supposed to work as an anti-skeptical theory that shows how we can have many justified beliefs without having justification for believing that certain skeptical hypotheses do not hold.

Going on to the latter notion, dogmatism also holds that perceptual justification can be defeated by additional evidence. This is captured by its claim that perceptual justification is prima facie: if you have strong evidence against *p*, or if a reliable source tells you that you are currently experiencing a visual illusion, then even though your perceptual experience that *p* provides you with prima facie justification for the belief that *p*, your belief that *p* will still not be ultima facie justified.

The dogmatist thesis is compatible with several views about what it is about perceptual experience in virtue of which the thesis is true. For instance, a reliabilist dogmatist could hold that the dogmatist thesis is (contingently) true because the process from perceptual experience to perceptual belief happens to be generally reliable. In contrast, a *phenomenalist* dogmatist holds that the thesis is true (at least in part) because of the intrinsic phenomenal properties of perceptual experience. It is the latter phenomenalist notion of dogmatism that will be our main concern in this chapter, although I will briefly reflect on a reliabilist version of dogmatism in Sect. 3.5. The phenomenalist version of dogmatism is also the one that its main proponents have been defending, which will become abundantly clear in Sect. 3.4.

Furthermore, the way the dogmatist thesis is usually phrased leaves open whether it subscribes to experientialism or not. Having an experience could be logically sufficient for having justification simply because it is correlated with a reliable process that non-evidentially justifies perceptual beliefs. However, it should be clear that our concern here is a variant of dogmatism that subscribes to experientialism: a perceptual experience that *p* is sufficient for justification of the belief that *p* because it evidentially justifies the belief that *p*.[3]

One might wonder, though, whether phenomenal dogmatists like Pryor, Huemer and Chudnoff are really committed to experientialism. Pryor explicitly states that "[...] it would be misleading to call these experiences your "evidence" for believing

[2]Note that I do not present the dogmatist thesis as "*necessarily*, if it perceptually seems to you that *p*, then you thereby possess some prima facie justification for believing that *p*". Although some dogmatists explicitly subscribe to this latter thesis (e.g., Tucker 2010), the necessity claim is left out in the original formulation of the thesis by Pryor (2000).

[3]Note that reliabilist dogmatist experientialism is also an option, Comesaña's (2010) evidentialist reliabilism appears compatible with such a view.

p" (2000, p. 519). However, Pryor just wants to convey here that a subject does not himself reason from the premise "it seems to me that *p*" to the conclusion "*p*". A subject need not have any beliefs about the relation between an experience and the perceptual belief it justifies, even though it is necessary that there is a specific relation between the two: they need to have the same content. That the experience still works as an evidential justifier is suggested by the following comparison Pryor makes:

> Compare: when you have a justified belief that *p* & *q*, you are thereby also justified in believing *p*. But this justification for believing *p* does not rest on any awareness you may have of the fact that you have a justified belief that *p* & *q*. You do not need to be able to appeal to the fact that you have a justified belief that *p* & *q* as a premise. The mere having of a justified belief that *p* & *q* is enough for your justification for believing *p* to be in place.
> (Pryor 2000, p. 519)

Just as you do not need to reason from the premise "I justifiably belief that *p* & *q*" to the conclusion "*p*", you also do not need to reason from the premise " it seems to me that *p*" to the conclusion "*p*". All you need for justification of your belief that *p* is to *have* the experience that *p*, or to have the justified belief that *p* & *q*. But the latter case of justification presumably has to do with the fact that *p* is deducible from the proposition that *p* & *q*, which is a sure sign of evidential justification.

Moreover, the relation between perceptual experience and belief is supposed to be rational, not merely causal. Having a perceptual experience with the content that *p* is supposed to make it rational for the subject to believe that *p*, and that is why the subject has prima facie justification for the belief that *p*. This comes out more clearly when one considers the notion of well-foundedness, or doxastic justification, instead of the notion of propositional justification that we have been working with. Suppose that a subject has the perceptual experience that there is a tomato in front of him. In virtue of having this experience, the subject now has propositional justification for the belief that there is a tomato in front of him. However, if the subject believes that there is a tomato in front of him merely because he experienced something red, then his belief will not be well-founded despite being propositionally justified. The problem is that the subject did not *base* his belief on the *evidence* that made it rational to hold it, i.e., his experience that there was a *tomato* in front of him.

So I think it's fair to say that phenomenalist dogmatist experientialism (from now on simply phrased as 'dogmatism') is the position that has been defended by philosophers like Pryor, Huemer and Chudnoff. And for good reasons too. In contrast with theories that require perceptual beliefs to be evidentially justified by other beliefs (about the reliability of one's own perceptual processes for instance), dogmatism merely requires that experience itself evidentially justifies perceptual belief. Together with a liberal view about the content of experience, which holds that experiences can represent kind properties like *being a table*, or *being a snake*, this makes it easy to explain how we are justified in all of our ordinary perceptual beliefs (I will go into more detail about this matter in the next section). Moreover, this justification might even be available to animals and children, as long as they are somehow capable of basing their beliefs on their experiences in the right way.[4]

The phenomenalist aspect of dogmatism further makes it into a full-fledged internalist view about justification, in the sense that we have access to those factors in virtue of which we have justification for our perceptual beliefs, viz. our perceptual experiences.[5] This provides it, in contrast with reliabilist versions of dogmatism, with the means to accommodate the two important internalist intuitions introduced in the previous chapter: the New Evil Demon Intuition and the Blindsight Intuition. The first intuition can be explained by pointing to the fact that subjects under the influence of an evil demon still have the same experiences, and therefore, the same justification as us. The second intuition can be explained by pointing to the fact that blindsighted subjects lack experience, and therefore lack the paradigmatic way of gaining justification for their perceptual beliefs.[6] As dogmatism is able to accommodate both intuitions about justification, and does not fall prey to an obvious hyper-intellectualization objection, it certainly appears to be a strong contender for an adequate account of perceptual justification.

3.3 The Nature of Perceptual Experience

3.3.1 Two Problems for Dogmatism

So far we have been going along with the dogmatist thesis as if it is intuitively clear what perceptual experiences are supposed to be. Most dogmatists stress that perceptual experiences, or perceptual seemings, are distinct from beliefs. For instance, when someone who is familiar with the Müller-Lyer illusion is looking at the Müller-Lyer lines, he will have the perceptual experience that the lines are unequal, but he will not believe that the lines are unequal. This provides some reason to think that perceptual experiences are distinct from beliefs.

---

[4]This last condition might actually be problematic. For instance, Huemer says the following about the basing relation: "when one apprehension, B, is based on another, A, A causes B because A (apparently) logically supports B" (2001, p. 56). This suggests that a subject needs to recognize, or think he recognizes, that an experience supports a certain belief for his belief to be based on that experience. And this seems too demanding for unsophisticated agents.

[5]Note that access to the experiences should be spelled out here in terms of being conscious of the experiences rather than believing that one has the experiences. The latter would immediately lead to the hyper-intellectualization worry.

[6]Note that, given that dogmatism merely provides a sufficiency claim for justification, it is merely compatible with the Blindsight Intuition but does not imply that it is correct. However, the dogmatist could easily make perceptual experience necessary as well as sufficient for *perceptual* justification. The main reason why the general claim is not phrased in this way stems from the fact that there are, of course, different means by which beliefs can be justified (e.g., my belief that there is a laptop in front of me could also be inferentially justified).

The fact that perceptual experiences are distinct from beliefs also helps dogmatists to solve the problem of infinite regress that we encountered several times in the previous chapter. The idea is that all justified beliefs are based on some internal ground, and that at the bottom of these grounds one finds perceptual experience.

But even with this distinction between experience and belief in mind, there is a need for further clarification. Consider the following example about perceptual identification, raised by Richard Feldman and Earl Conee:

> A novice bird watcher and an expert are together looking for birds. They both get a good look at a bird in a nearby tree. (In order to avoid irrelevant complexities, assume that their visual presentations are exactly alike.) Upon seeing the bird, the expert immediately knows that it is a woodpecker. The expert has fully reasonable beliefs about what woodpeckers look like. The novice has no good reason to believe that it is a woodpecker and is not justified in believing that it is.
>
> (Feldman and Conee 2001, pp. 3–4)

Feldman and Conee explain the difference in justification on the basis of the fact that the expert, but not the novice, knows the look of a woodpecker. Yet this goes against the intuitive idea that the expert has *immediate* justification for his belief. In any case, *if* the expert has immediate justification for his belief that the bird is a woodpecker, then there is a problem for dogmatism. For ex hypothesi both expert and novice have visual presentations that are exactly alike, and therefore at least appear to have the same perceptual experiences. But if they have the same perceptual experiences, then they should have immediate justification for the same beliefs. Yet only the expert has immediate justification for the belief that the bird is a woodpecker. So dogmatism needs to give an account of perceptual experience that explains why the expert, but not the novice, has the perceptual experience that the bird is a woodpecker.

Another problematic case for dogmatism is that of the speckled hen. Ernest Sosa presents this as a problem for classical foundationalism, but it is equally problematic for dogmatism:

> If the classical foundationalist wishes to have a theory and not just a promissory note, he needs to tell us which sorts of features of our states of consciousness are the epistemically effective ones, the ones by corresponding to which specifically do our basic beliefs acquire epistemically foundational status. Having a visual image with forty-eight speckles seems not to qualify, whereas having a visual image with three speckles may (at least when they are large and separate enough). What is the relevant difference?
>
> (BonJour and Sosa 2003, p. 121)

The problem for dogmatism is that we can have immediate justification for the belief that the hen has 3 speckles on the basis of the 3-speckled visual image, but that we cannot have immediate justification for the belief that the hen has 48 speckles on the basis of the 48-speckled visual image.[7] What dogmatism needs is a good

[7] I'll admit that both for Feldman and Conee's example and Sosa's, it's not entirely clear what is meant with respectively 'visual presentation' and 'visual image'. But the cases can also be understood simply as a challenge for dogmatism to explain the difference in epistemic status by elucidating the notion of perceptual experience.

explanation that shows why the perceptual experiences are different in these two cases, otherwise it is left with an unexplainable difference in immediately justified beliefs. Again, the lesson is that we need to have a more detailed account of what perceptual experiences are exactly.

3.3.2 Distinguishing Sensations and Seemings

In the face of the previous problems, several authors have given a similar account of perceptual experience by distinguishing between perceptual seemings and perceptual sensations (Tucker 2010; Bergmann 2013; Brogaard 2013b). The epistemic work is supposedly done by the perceptual seemings rather than the perceptual sensations. This would provide a good reason to present dogmatism in the way Chudnoff did, as the view that "if it perceptually seems to you that *p*, then you thereby possess some prima facie justification for believing that *p*" (Chudnoff 2011, p. 314).[8] However, the notion of a perceptual seeming is itself somewhat unclear, even if it is elucidated by contrasting it with both sensations and beliefs. I therefore interpret authors like Tucker, Bergmann and Brogaard as talking about high-level *percepts* when they talk about perceptual seemings, which should be distinguished from low-level sensations.[9]

The relevant distinction between sensations and percepts is explained by Jack Lyons (2009) in the following way:

> I want to invoke a well-known distinction here, which, although in fact a somewhat outdated oversimplification of perceptual experience, allows us to pick out two ends of a spectrum and will help to bring some order to the discussion of nondoxastic experience. Most introductory psychology textbooks draw a well-known distinction between sensation and perception. As I will use the terms, sensation is the raw, direct experiential consequence of the stimulation of the sense organs, while perception involves processing of information. The sensation is what a Lockean *tabula rasa* would experience, while the perception [or the percept] is the result of the mind's unconscious and involuntary attempt to make sense of the world. Perception transforms James's "blooming buzzing confusion" of sensation into meaningful experience.[10]
>
> (Lyons 2009, p. 42)

To clarify the distinction, Lyons appeals to several examples. For instance, when one looks at the Necker cube and switches perspective, what changes is the percept

[8] Note, though, that Chudnoff himself does *not* distinguish between sensations and seemings, but thinks that experiences and seemings are identical (Chudnoff 2014).

[9] An exception is Matthew McGrath (2013a), who appears to take *both* sensations and percepts to be seemings. My account of perceptual seemings will have negative consequences for the theory he develops in both McGrath (2013a,b).

[10] Note that Lyons remarks the distinction is a "somewhat outdated oversimplification". Lyons merely uses the distinction to argue against another position and does not seem to need it to support his own.

even though the sensation remains the same (Lyons 2009, p. 43). Something similar seems to be applicable to other ambiguous figures, such as the duck-rabbit. When one switches perspective from duck to rabbit, the sensation remains the same but the percept changes. To take a more natural example, Lyons remarks that a lump of coal in bright light and a snowball in dim light reflect the same amount of light, thereby causing the same sensations. Yet because the visual system takes into account the ambient lighting conditions, the percepts of the two objects are still different. So there is a sense in which the two objects look the same (because they produce similar sensations) and also a sense in which they look different (because they cause different percepts) (Lyons 2009, pp. 43–4).

We can now distill several ideas about sensations and percepts.

1. Sensations and percepts are distinct conscious states. This is reflected in Lyons' claim that sensation is the "raw, direct *experiential* consequence" of sensory stimulation, while a percept is a transformation of this sensation into a "meaningful *experience*".
2. Sensations are experiential states that supervene on *early* stages of perceptual processing, while percepts are experiential states that supervene on *late* stages of perceptual processing. At the least, sensations arise *before* percepts in the perceptual process. This comes out clearly in the idea that sensations are supposed to be *direct* experiential consequences of sensory stimulation, while percepts are the result of some sort of transformation of sensation.
3. Sensations only represent, if anything, low-level properties like shapes and colors, while percepts can also represent high-level properties like 'being a duck'. It is this aspect of sensations and percepts that explains why something changes and something stays the same when one switches perspective while looking at an ambiguous figure: the same low-level properties are still represented by the same sensation, but there is a change in which high-level properties are represented by the new percept.

Several authors that have defended the importance of seemings for justification appear to rely on this distinction between sensations and percepts, although they do not explicitly identify seemings with percepts. One explanation of this is that these authors are interested in dogmatism in its full generality, instead of a mere *perceptual* dogmatism. Nevertheless, I think that from what they claim about seemings in relation to perception, it is very plausible that they have something like percepts in mind for the case of perception. For instance, Michael Bergmann claims the following:

> I think that the view that sensory experience has propositional content is a mistake that stems from the temptation to conflate the experience and the seeming. It's the seeming that has the propositional content, not the sensory experience. But the sensory experience naturally produces that seeming and is constantly conjoined with it, so it is tempting (though mistaken) to assign the seeming's content to the experience.
>
> (Bergmann 2013, p. 167, n. 9)

All three aspects of sensations and percepts are part of Bergmann's distinction between sensory experience and seeming: both are conscious states, the former is prior to and produces the latter, and the former has no propositional content while the latter does.[11]

Chris Tucker can also be taken to accept something much like the above sensation-percept distinction:

> It is commonly assumed that a sensation is a special kind of seeming [...], but I argue that this [is] incorrect. Sensations, like seemings, are experiences, and it is plausible that at least some of them have content. I have a visual sensation when I look at my dog. It is the mental "picture" or visual image of a little white creature wearing a blue halter. I have an auditory sensation when I hear my dog barking. It is the mental "sound" of the bark, a mental phenomenon that causes me great irritation. At the very least, there are also tactile, olfactory, gustatory and perhaps even proprioceptive sensations.
>
> (Tucker 2010, p. 530)

Tucker accepts both (1) and (3): sensations are experiences that represent low-level properties (in the case of vision, they are mental "pictures"), while seemings are experiences that can also represent high-level properties.[12] Tucker does not accept (2) because he thinks that seemings typically *accompany* sensations, where accompanying is understood as occurring at the *same* time (Tucker 2010, p. 543, n. 12). However, no motivation is given for this peculiar view.

Finally, Berit Brogaard distinguishes between perceptual states (or perceptual experiences) and states of seemings:

> My perceptual experience may represent 17 people in the seminar room even [if] it doesn't visually seem to me that there are (exactly) 17 people in the seminar room. So perceptual states and states of seemings are different kinds of mental states. One can now say that while the expert and the novice have perceptual experiences with the same experiential content, they are in different states of seemings concerning the tree. The overall difference in phenomenology between the expert and the novice stems from a difference in the phenomenology of the two distinct states of seeming. To the expert, the tree visually seems to be an Elm, it visually seems familiar and it visually seems different from the nearly identical neighboring tree which is not an Elm. To the novice, the tree just seems to be a tree, it doesn't seem familiar and it seems exactly similar to the neighboring tree.
>
> (Brogaard 2013a, p. 37)

Brogaard accepts (1) and (3)—only seemings can represent properties like 'being an Elm' although both seemings and perceptual experiences have an impact on phenomenology—and she explicitly claims that seemings are interpretations of experiences (Brogaard 2013b, p. 278), thereby accepting (2). So Brogaard also appears to rely on the sensation-percept distinction although she does not frame the distinction in those terms.

The quote from Brogaard already makes clear how the distinction between sensation and percept can be used to respond to the above problem of explaining the epistemic difference between perceptual judgments of novices and experts. Even though a novice bird watcher and an expert both have the same *sensations* when looking at a woodpecker, they have different *percepts*: the percept of the novice

[11]Note the "if anything" in condition (3).

[12]This becomes perfectly clear in his discussion of associative agnosia (Sect. 3.3.3.2).

merely represents the object under consideration as a bird, without representing it as a specific bird, while the percept of the expert represents the object as a woodpecker. If we now assume that perceptual seemings should be identified with percepts, this explains why, even though novice and expert have the same "mental picture" of an object, the latter is immediately justified in a more specific perceptual belief than the former according to dogmatism.[13] It perceptually seems to the novice that there is a *bird* in front of him because his percept represents that there is a bird in front of him, while it perceptually seems to the expert that there is a *woodpecker* in front of him because his percept represents that there is a woodpecker in front of him. This is so even though both novice and expert have the same sensation of the bird.

In a similar way, the distinction between sensation and percept can be used to respond to the speckled hen problem (Tucker 2010; Brogaard 2013b). When a human subject looks at a hen with three speckles, he has a sensation that represents a three-speckled hen and a percept which represents that the hen has three speckles. When a human subject looks at a hen with 48 speckles, he has a sensation that represents a 48 speckled hen, but he has a percept which only represents that, say, the hen has between 42 and 55 speckles (given that adults can accurately distinguish numerosities with a 7:8 ratio) (Barth et al. 2003). This is just a contingent fact about how human subjects represent large numbers in perception. Assuming again that perceptual seemings should be identified with percepts, this explains why a hen with 3 speckles perceptually seems to have 3 speckles, while a hen with 48 speckles does not perceptually seem to have 48 speckles, even though in both cases the hen and its speckles are represented in a similar way by the sensation.

Now why do I insist on identifying perceptual seemings with percepts when the mentioned authors do not explicitly support this identification? I think that without this identification, the responses to the problems simply lack plausibility. Take Tucker's response to the problem of the speckled hen as an example:

> In normal humans, a visual image of a three-speckled hen is accompanied by a seeming that the hen has three speckles, and the visual image of the forty-eight-speckled hen is not accompanied by a seeming that the hen has forty-eight speckles. This difference in the way things seem is what explains the difference in justification.
>
> (Tucker 2010, p. 535)

Without making more explicit what these seemings are, this response will simply strike anyone as ad hoc.[14] By identifying seemings with percepts, it becomes clearer what they amount to and less ad hoc to invoke them as separate entities from sensations. If proponents of dogmatism want to maintain a distinction between perceptual seemings and sensations without identifying seemings with percepts, then they should provide a better explanation of how these seemings are realized and what their connection is to sensations.

---

[13]Tucker provides a similar response to the problem of non-inferential perceptual identification (Tucker 2010, p. 538).

[14]Note that Tucker does provide separate motivations for the sensation-seeming distinction. However, as I argue in the next section, these motivations are not very strong.

In the following section I will take for granted that perceptual seemings should be identified with percepts, and thereby take the sensation-seeming distinction to be identical with the sensation-percept distinction.

3.3.3 Against the Sensation-Seeming Distinction

Although I think that there is something right about the previous response to the problem of novice vs. expert perceptual identification and the problem of the speckled hen, I also think that there is something wrong about it. What's right is that perceptual seemings should not be identified with low-level representations, but what's wrong is that there is no good reason to assume that sensations and seemings exist as separate conscious states. Instead, perceptual experiences should simply be taken to incorporate aspects of *both* sensations and seemings. To argue this point, I will start by undermining some of the motivations for the distinction between sensations and seemings presented by Tucker (2010), and later present some reasons to support my own contention about perceptual experience.

According to Tucker, the distinction between sensation and seeming can be supported by two phenomena from empirical psychology: the phenomenon of blindsight and the phenomenon of associative agnosia. Recall that blindsighted subjects have a damaged visual cortex and strongly claim that they cannot see anything in a certain part of their visual field. However, when they are forced to guess about, say, the identity of a certain visual figure or the location of a certain object in their blind spot, they perform much better than chance—even to their own surprise. Appealing to a distinction between sensation and seeming might explain what is going on here: "the subjects' "blind spots" are regions in their visual fields that lack visual imagery. Nonetheless, the mechanisms that produce seemings function well enough to provide information about the region of the environment that corresponds to the subjects' blindspots" (Tucker 2010, p. 530). So blindsighted subjects are supposed to have seemings about what is going on in their blind field, even though they lack sensations in that area.

Going on to the second phenomenon, patients with associative agnosia are able to accurately copy pictures of common objects (or draw common objects they are presented with), like pens, rings, scissors, etc., but are unable to recognize these objects. On Tucker's account, what is going on here is that "subjects can draw the object because their visual imagery (i.e., their sensation) is intact. They cannot recognize it because they fail to have the seemings which would enable them to do so" (Tucker 2010, p. 531). So the idea here is that subjects with associative agnosia have sensations of the objects in front of them, even though they lack certain seemings about them.

If these explanations are correct, then there is good reason to accept a distinction between sensation and seeming: there would then be subjects who have the former but not the latter (associative agnosia) as well as subjects who have the latter but not the former (blindsight). However, both empirical phenomena actually fall short of strongly motivating a distinction between two separate conscious states. To do so, the explanations of these phenomena should (a) provide a good description and explanation of all the relevant data, and, (b) should be preferable over rival explanations. I will argue that Tucker's explanation of blindsight fails to meet requirement (a), while his explanation of associative agnosia fails to meet requirement (b).

3.3.3.1 Blindsight

Let's begin with the case of blindsight. Tucker makes the following two claims about this phenomenon:

1. A blindsighted subject has no sensations ("visual imagery") in a certain region of his visual field.
2. A blindsighted subject has seemings about that part of the environment which is located in his blind field.

Although the first claim might be attacked on the grounds that it is not clear that a blindsighted subject has *no* sensations in a specific region of his visual field (cf. Overgaard 2011), the theory can easily be adapted by claiming that the sensations in that region are strongly impoverished. The real problem has to do with the second claim, which is at odds with the fact that the blindsighted subjects are genuinely surprised (as Tucker acknowledges) about the correctness of their own answers. Given that a seeming is a conscious experience with a phenomenology that "makes it feel as though the seeming is "recommending" its propositional content as true" (Tucker 2010, p. 530), one would not expect blindsighted subjects to be surprised about the correctness of their guesses. Why be surprised at the correctness of your guess if you have an experience which presents the content of your guess as being true?

Tucker is aware of this tension and claims that "subjects are not confident about their guesses because the relevant seemings are very weak and the subjects understandably assume that, in the circumstances, they do not have any reliable access to what is taking place in their blindspot" (Tucker 2010, p. 530–1). In this reply, Tucker appeals to his view of seemings as being weak or strong depending on how strongly they assert their content (Tucker 2010, p. 530). However, even accepting that there are both weak and strong seemings, there are still several problems with this response. First of all, there is no clear relation between the strength of a seeming and the degree of confidence a subject has in the content of that seeming. Having a strong seeming is compatible with having a low degree of confidence in its content (just think of someone looking at the Müller-Lyer lines while being familiar with their illusory nature), and having a weak seeming is compatible with having a high degree of confidence in its content (e.g., a mathematician confident in a very counter-intuitive but proven mathematical theorem).

Second, even if there was a relation between the strength of a seeming and the degree of confidence in its content, then the first part of Tucker's remark—that blindsighted subjects are not confident about their guesses because the seemings are very weak—is still besides the point. Suppose that blindsighted subjects are not confident in their guesses because the contents of these guesses are only weakly presented as true by their seemings. Then they should still not be as strongly surprised as they are if what weakly seems to be true actually is true. To compare: suppose I'm a candidate in a game show where I can choose to either win a certain amount of money, or, try to double the money by answering the current multiple choice question before me. If C weakly seems to me to be the right answer, then I will not be surprised if C actually turns out to be correct. This is so even if I am not confident enough to actually risk my money by answering the question.

The second part of Tucker's reply is more to the point. If blindsighted subjects believe that they do not have any reliable access to what is taking place in their blind spot, then this explains why they are surprised when they are correct in their forced guesses. But here the difficulty lies in explaining why subjects would "understandably" have this belief. According to the dogmatist, beliefs are typically caused (and justified) by seemings (Huemer 2001, pp. 55–7), and the relevant seemings here should cause subjects to believe that such-and-such is taking place in their blind field, *not* that they do not have any access to what is going on in the blind field. The most plausible route around this problem would suggest that there is also another, stronger seeming at play, which has the content that the subject does not have any reliable access to what is taking place in his blind field. But where does this extra seeming come from? It is strange to suggest that one mechanism simultaneously produces a strong seeming that there is no reliable access to what is going on in the blind field, and a weak seeming that such-and-such is going on in the blind field. So maybe there are two different mechanisms at play, one of which produces the strong seeming that there is no reliable access to what goes on in the blind field, and another which produces a weak seeming that such-and-such is going on in the blind field. The fact that the first strong seeming is incorrect elicits the surprise of the subjects.

The picture that emerges is one according to which blindsighted subjects have multiple conflicting seemings, one strong seeming with the content that the subject does not have any reliable access to what is going on in his blind field, and one weak seeming with the content that such-and-such is going on in the blind field. If forced to guess, blindsighted subject base their guess on the weak seeming, but they have no confidence in their guess because they also have the conflicting strong seeming. However, this picture is not in accordance with the fact that blindsighted subjects themselves are not aware of any conflict between seemings. Nor are they aware of basing their guess on any weak seeming at all. Tucker gets the phenomenology wrong by suggesting that subjects have seemings which lead them to the guesses they make: it is to the subjects as if they are *randomly* guessing—as Tucker himself remarks (2010, p. 530).

Couldn't the subjects have based their guesses on unconscious seemings? Perhaps, but then it is no longer clear what seemings are supposed to be, as they were introduced as conscious experiences with specific phenomenologies. Moreover, this would have the devastating consequence that dogmatism could no longer be combined with access internalism: it's perceptually seeming to S that p would no longer be accessible to the subject in cases in which the perceptual seeming is unconscious.

The phenomenon of blindsight thus does not lend any support to a theory which distinguishes between the conscious states of sensation and seeming. It is also not inconsistent with such a theory, since a proponent could maintain that neither sensations nor seemings are present for a certain region of a blindsighted subject's visual field. However, given that Tucker wanted to use blindsight as a motivation for endorsing the theory, this example fails. This puts extra pressure on the one remaining case that is supposed to motivate the theory, namely, the phenomenon of associative agnosia.

3.3.3.2 Associative Agnosia

Tucker makes the following two claims about the phenomenon of associative agnosia:

1. Subjects with associative agnosia have intact sensations of the objects they are unable to recognize.
2. Subjects with associative agnosia lack relevant seemings about the objects they are unable to recognize.

The first claim is supported by the fact that subjects with associative agnosia are able to accurately draw the objects they are unable to recognize, while the second claim is supported by the fact that these subjects cannot identify the objects. This certainly appears to be a valid explanation of what is going on in associative agnosia, but if it is supposed to motivate the distinction between sensations and seemings, then it should be preferable over relevant rival explanations. One such rival is exactly one which does not accept sensations and seemings as separate conscious states, but holds instead that there is just one conscious state, *perceptual experience*, which incorporates aspects from both sensations and seemings.[15] The rival explanation would then be that the perceptual experiences of patients with associative agnosia are simply impaired because of a certain deficit of their perceptual system.

Tucker seems to be aware of such a rival explanation and considers it in a footnote, although he expresses it differently by taking the opponent to hold that sensations also represent high-level properties, thereby making seemings redundant:

> Do the sensations of agnosia patients represent the key as a key and not merely, say, as an object with some particular shape and color? If they do, then my explanation of these agnosia cases may depend on some further explanation for why recognitional capacities are determined by seemings rather than sensations. Alternatively, I might argue that no sensation, not just those of agnosia patients, can represent properties like being a key.
>
> (Tucker 2010, p. 543, n. 11)

[15]In fact, the phenomenon of associative agnosia seems to have *also* been used by Tim Bayne (2009) to defend such a rival view. But I think that Brogaard (2013a) is correct in her contention that associative agnosia also does not motivate this view over the alternative.

This response will not do considering the current dialectics. If the example of associative agnosia is meant to make the distinction between sensation and seeming empirically plausible, then we should be presented with an argument (instead of a possible argumentative route) that explains why this account is to be preferred to alternative explanations of the phenomenon. Given that Tucker does not provide such an argument, the phenomenon of associative agnosia does not motivate a distinction between sensation and seeming.

3.3.4 High-Level Perceptual Experience

Given that both blindsight and associative agnosia do not motivate a distinction between sensation and seeming, what, then, can be maintained about the distinction? In fact, I think there is sufficient reason to be skeptical about all three aspects of the sensation-seeming distinction. With regard to (1), there is the phenomenological fact that, when we are perceiving, we are not usually aware of two separate conscious states which then lead to perceptual belief—in the usual case we are not even aware of a difference between perceptual experience and perceptual belief.[16] So the thesis that perceptual beliefs arise because of a combination of sensation and seeming is not phenomenologically supported. Of course, such introspective evidence is highly defeasible: if we have strong empirical evidence for the claim that there are two distinct conscious states active in perception, then this trumps the introspective evidence (as introspection often turns out to be unreliable (e.g. Schwitzgebel 2008)). However, such evidence has so far not been presented, precisely because blindsighted subjects do *not* have *conscious* perceptual seemings.

With regard to (2), there is reason to assume that late stages in perceptual processing also influence early stages in perceptual processing. There are so-called 'back-projections' of information from brain areas used in late stages of visual processing to brain areas used in early stages of visual processing that clearly influence perceptual awareness (Pascual-Leone and Walsh 2001; Silvanto et al. 2005). Jesse Prinz explains the function of these back-projections as follows:

> In vision, the objects around us are often occluded, poorly illuminated, insufficiently foveated, or moving too quickly to be adequately perceived. We take in enough information to make good guesses about object identity, but doing so often requires that we fill in information that has been lost. To do this, we use available information to call up less degraded perceptual representations from memory, which can be back-projected into earlier visual areas to enhance the signal.
>
> (Prinz 2006, p. 455)

If this is correct, then this also has some import for aspect (3) of the sensation-seeming distinction. If the representations at early stages of perceptual processing

[16] Lyons (2009, p. 71) even suggests that experience and belief might be token-identical in the usual case.

are partly formed on the basis of back-projections from late stages of perception, then it becomes at least a relevant possibility that these representations also represent high-level properties due to the influence of these back-projections (Prinz 2006). Other evidence against (3) consists in the fact that some high-level properties can be perceived without attention and at very high speeds (Fish 2013), which one would not expect if high-level properties were only represented at *later* stages of perceptual processing.

Taken together, this gives us sufficient reason to shift the burden of proof to proponents of the sensation-seeming distinction, and to seriously consider the alternative that perceptual experience is a conscious state that supervenes on all levels of perceptual processing, incorporating aspects from both sensations and seemings.[17] This also has large implications for the debate focusing on the question of whether high-level properties can be represented in perception.[18] Given that I have argued that the default position about perceptual experience should be that it also supervenes on stages of perceptual processing in which high-level properties are represented, this strongly suggests that, at least as a default position, high-level properties should be taken to be part of the representational content of perceptual experience.[19] This will allow dogmatism to give the following answers to the two earlier problems of novice vs. expert identification and the problem of the speckled hen.

In the case of the novice and expert bird watcher, it simply is not true that both have the same perceptual experience. The novice has a perceptual experience which represents that there is a bird in front of him, while the expert has an experience which represents that there is a woodpecker in front of him. That is why it perceptually seems to the novice that there is a bird in front of him, while it perceptually seems to the expert that there is a woodpecker in front of him. Similarly, when we look at a 3-speckled hen our perceptual experience represents that the hen has 3 speckles, but when we look at a 48-speckled hen our perceptual experience merely represents that the hen has, say, between 42 and 55 speckles.[20] That is why the 3-speckled hen perceptually seems to have 3 speckles, while the 48-speckled hen perceptually seems to have between 42 and 55 speckles.

Although this explanation of the problems seems to resemble the earlier explanation that made use of a distinction between sensation and seeming, it has some clear benefits. First, it is a simpler response to the problems as it does not posit an

[17]Note that Lyons' sympathies also lie with such a view despite his argumentative use of the sensation-percept distinction: "experience [...] is a matter of the introspectible features of a host of different representational states" (Lyons 2009, p. 48).

[18]Susanna Siegel (2006), Tim Bayne (2009), Farid Masrour (2011), and William Fish (2013) are in favor of the idea that high-level properties are represented in perception, while e.g. Michael Tye (2000), Berit Brogaard (2013a), and Indrek Reiland (2014) are against it.

[19]Although this does not mean that the phenomenology of perceiving these high-level properties is similar to the phenomenology of seeing low-level properties. As Prinz (2006) puts it, there might not be any "poodle-qualia" just as there are red-qualia.

[20]Michael Tye (2009) gives a similar response to the problem of the speckled hen.

extra theoretical entity in the form of a sensation or seeming. Second, the proposed account appears to be empirically better grounded than the one invoking a sensation-seeming distinction.

This shows that some of the problems for dogmatism can be adequately overcome by developing a more precise and empirically grounded account of the nature of perceptual experience. Fortunately, some dogmatists have already explicitly committed themselves to the thesis that perceptual experiences and perceptual seemings are the same. According to Chudnoff (2014), without identifying experiences and seemings, one allows for the counter-intuitive possibility that there could be justifications because of *mere* seemings, states that present *p* as true without doing it in any particular way (e.g., visually, introspectively, intellectually). Given that we would not normally accept such mere seemings as justifying sources, we should also not accept a view which distinguishes seemings from experiences.[21]

However, even with this improved account of perceptual experiences and perceptual seemings in hand, a specific problem for dogmatism remains. Although this problem is also partly related to the nature of perceptual experience, it is unlikely that it can be solved in a similar manner as before. The reason for this is, as was the case with evidentialism, that the problem appears to be of a more conceptual rather than empirical nature. It is to this problem that we will now turn.

3.4 The Distinctiveness Problem

Recall that dogmatism claims that having a perceptual experience that *p* is sufficient (evidence) to have immediate prima facie justification for the belief that *p*. The last section has looked in more detail at what is meant by having a perceptual experience that *p*, and concluded that they should be seen as conscious states with high-level representational content. This section will assume that this account of perceptual experience is correct, but still questions whether it provides dogmatism with the means to support its sufficiency claim.

According to dogmatism, perceptual experience can evidentially justify belief without requiring any justification itself, or requiring any other justified beliefs about e.g., the denial of sceptical hypotheses. In a sense, this was exactly what we wanted, because this prevents both the problem of an infinite regress of justification (because experiences are not in need of justification themselves), and the problem of hyper-intellectualization (because no other justified beliefs are presupposed). However, it does raise the question of *how* perceptual experience is able to perform this amazing feat. This question becomes especially pressing when one realizes that some paradigm examples of propositional states do require justification themselves

[21]Note that Chudnoff (2014) also believes that dogmatists have a different way of responding to the speckled hen (and novice vs. expert) problem, but I will get to that in Sect. 3.5.

before they can evidentially justify beliefs, while other paradigm examples of propositional states cannot evidentially justify any beliefs at all.

Consider an obvious case of evidential justification in which the belief that *p*, together with the belief that if *p* then *q*, justifies the conclusion that *q*. The conclusion will only be justified if the premise beliefs themselves are both justified. This shows that some states with propositional content can only evidentially justify beliefs if they are themselves justified.

On the other hand, consider a case in which I imagine that *p*, or strongly desire that *p*. These propositional states will not be able to evidentially justify a belief that *p* even in the case where my imagination or desire is so strong that I actually start believing that *p*. Note that imagination might be able to justify beliefs in other cases, but it certainly is not able to justify a belief that *p* solely on the basis of imagining that *p*.

According to Lyons (2009, pp. 74–5), *both* these cases can be explained from the supposition that all evidentially justified beliefs derive their justification from whatever justified the evidence in the first place. If this is correct, then all evidential justifiers must themselves be justified before they can confer any justification, because there would not be any justification to confer otherwise. Since only beliefs appear to be states with propositional content that are capable of being justified, Lyons concludes that only beliefs can evidentially justify other beliefs.

Even if one does not agree with Lyons' specific explanation of these facts concerning evidential justification, he has still succeeded in presenting a strong challenge for dogmatism. Dogmatism has to provide an alternative account of the difference between belief, experience, desire, and imagination that explains why the first two are possible evidential justifiers, while the latter two are not. Moreover, it has to give an account of what is so special about experience as an evidential justifier that enables it to provide immediate prima facie justification without being itself justified. Call this the *Distinctiveness Problem* for dogmatism. Without an answer to the Distinctiveness Problem, dogmatism would remain an ad hoc view of perceptual justification.

3.4.1 Phenomenalist Answers

3.4.1.1 A Feeling of Seeming to Ascertain That p

Most proponents of dogmatism provide at least some suggestion as to what it is about perceptual experience that gives it the power to evidentially justify beliefs without being in need of justification itself. Usually these suggestions appeal to some special phenomenological property that is distinctive of perceptual experience and supposed to be epistemically relevant. For instance, Pryor gives the following answer to the Distinctiveness Problem:

> I think there's a distinctive phenomenology: the feeling of *seeming to ascertain* that a given proposition is true. This is present when the way a mental episode represents its content makes it feel as though, *by* enjoying that episode, you can *thereby just tell* that content

> obtains. [...] When you daydream or exercise your visual imagination, you represent propositions [...], but it does not feel as though you can thereby just tell that those propositions are true.
>
> (Pryor 2004, p. 357)

According to Pryor, the special epistemic status of perceptual experience is due to the connected feeling of seeming to ascertain that a given proposition is true. But one problem with this suggestion is that it's not clear whether the feeling of seeming to ascertain that a given proposition is true necessarily has to do with the way a mental episode, like perception, represents its content. Couldn't the feeling arise because of a separate belief about the mental episode? One candidate for such a belief would be the belief that the mental episode is veridical: if someone believes that he is veridically perceiving that *p* (even if he is, say, imagining that *p*), then it's no wonder that he also has a feeling of seeming to ascertain that *p* is true.[22]

Now one thing one might claim in response to this proposal is that, in the case under consideration, the higher-order belief that one is veridically perceiving that *p* actually causes or perhaps even constitutes a perceptual experience that *p*. According to this response, if one imagines that *p* and simultaneously believes that one is veridically perceiving that *p*, one simply undergoes a perceptual experience that *p*. However, there are two problems with this response. First, for it to fully work against the problem, one would have to suppose that higher-order beliefs of a certain sort are always sufficient to produce perceptual experiences. Otherwise we can just change the example to one where a subject does not undergo any sensory experience at all, yet still has the higher-order belief that he is veridically perceiving that *p* (thereby producing a feeling of seeming to ascertain that *p* is true). And it simply appears to be implausible that higher-order beliefs on their own are sufficient for producing perceptual experiences.

More importantly though, even if one did hold that the combination of imagination and higher-order belief was sufficient for perceptual experience, then there is a second problem left. Even though the seeming of feeling to ascertain that *p* would on this account be *distinctive* of a perceptual experience that *p* (almost as a matter of definition), it would no longer be able to support a plausible variant of dogmatism. On the current account, perceptual experiences with this phenomenological property would simply become too easy to obtain, thereby making perceptual justification too easy to obtain. Whenever one had the right kind of unjustified higher-order beliefs, one would have an experience capable of justifying beliefs, because it would have the right kind of phenomenology (i.e. a feeling of seeming to ascertain that *p* is true). This is simply unacceptable for any satisfying account of perceptual justification.

---

[22]Note that this does not deny Pryor's claim that it's not enough for prima facie justification to think you have the phenomenology (Pryor 2004, p. 357). I'm envisaging a situation in which a higher-order belief actually gives rise to the necessary phenomenology.

This problem is similar to the dogmatist problem with cognitive penetration.[23] If an unjustified belief or overly strong desire influences one's perceptual experience to such a degree that it becomes insensitive to how the environment actually is, then the perceptual experience no longer seems capable of providing justification for one's beliefs—even if it does have a specific kind of phenomenology. Perceptual justification will become too easy to obtain if it is just about having an experience with a specific content in combination with a feeling of seeming to ascertain that the content is true.

3.4.1.2 Presentational Phenomenology

Chudnoff (2011, p. 317) agrees that Pryor's account of the special phenomenology of perceptual experience is somewhat too underdeveloped to work. Chudnoff explicates Pryor's suggestion by appealing to an experience's *presentational phenomenology*, claiming that an experience has presentational phenomenology with respect to a proposition *p* "just in case they are experiences in which we both represent that *p* [...] and seem to be aware of an item that makes it the case that *p*" (Chudnoff 2011, p. 321). According to Chudnoff, it is an experience's presentational phenomenology that accounts for its justificatory force:

> I find it compelling that if you have an experience that not only represents your environment as being a certain way, but that is one in which you also seem to be aware of the very items in your environment in virtue of which it is the case that your environment is that way, then you thereby have some prima facie justification for believing that your environment is the way it appears to you to be.
>
> (Chudnoff 2011, p. 322).

The problem with this specific proposal is that it might not provide a satisfying answer to the Distinctiveness Problem because it fails to distinguish clearly between imagination and perception. This becomes prominent when one looks at the specific ways in which Chudnoff fills out presentational phenomenology for perception and intuition. For the case of perception, Chudnoff thinks that presentational phenomenology comes down to a combination of "seeming to fact-perceive that [*p*]" (2011, p. 319) and "seeming to be sensorily item-aware of an item [that makes it the case that *p*]" (ibid.). Similarly, "an intuition experience possesses presentational phenomenology when in it you both seem to fact-*intuit* that *p* and seem to be *intellectually* item-aware of an item that makes it the case that *p*" (Chudnoff 2011, p. 323). Then why not claim that a visualization has presentational phenomenology when in it you both seem to fact-*visualize* that *p* and seem to be *imaginatorily* item-aware of an item that makes it the case that *p*?

---

[23]See e.g., Markie (2005), Lyons (2011), Siegel (2012, 2013a), McGrath (2013a), and Tucker (2014) for more on this problem. Also see Ghijsen (2015) for more on the relation between the Distinctiveness Problem and the problem of cognitive penetration.

Although Chudnoff agrees that imagination can also have a sort of presentational phenomenology, he thinks that there is still an important difference between imagination and, e.g., perception:

> Unlike sensory, intellectual, or self-presentational awareness, objectual imagining is not a kind of awareness, or seeming awareness. That is, objectually imagining an *F* is not a way to seem to be aware of an *F*. On the face of it, if you seem to be aware of a tiger, for example, you represent the tiger as actual—it is really there. But you can objectually imagine a tiger without representing the tiger as actual.
>
> (Chudnoff 2012, p. 62)

The important difference that Chudnoff here points to has to do with representing something as actual. But this alone cannot distinguish imagination from perception. Although I agree with Chudnoff's point that you *can* objectually imagine a tiger without representing the tiger as actual, you can also objectually imagine a tiger *and* represent it as actual (e.g., by imagining that there actually is a tiger in front of you at this moment). But surely this imaginatory experience would then still not provide any prima facie justification for the belief that there is a tiger in front of you.

However, maybe we have to focus more on the intuitive idea that the way in which we are aware of the objects of imagination is different from the way in which we are aware of the objects of perception. When we perceptually experience objects, we phenomenally seem to stand in a different relation to those objects than when we merely imagine those objects.[24] But even though this is certainly true for the usual cases of perception and imagination, we can still question whether this is true for *all* cases. And the examples we can use are similar to the ones used above: for instance, what happens when one imagines that *p* while simultaneously believing that this experience makes one aware of a truth-maker for *p*? It seems that this combination of imagination and belief could give rise to presentational phenomenology, yet the full experience one then undergoes is still intuitively not capable of granting prima facie justification.

Now one might again respond by claiming that imagination and belief in this case give rise to a genuine perceptual experience, but the problem remains that this will not be a perceptual experience that is intuitively capable of granting prima facie justification. So instead of posing the Distinctiveness Problem as challenging dogmatists to provide a distinctive property of perceptual experience as a *type* that enables it to evidentially justify beliefs without being justified itself, one could also see it as challenging dogmatists to provide a distinctive property of those *token* mental states (like many perceptual experiences) that are intuitively able to evidentially justify beliefs without being justified themselves.

This way of reading the challenge also makes the problem of cognitive penetration (and perhaps the earlier problems of Sect. 3.3.1 as well) into a symptom

---

[24]Perhaps this is better captured by the novel way in which Chudnoff describes presentational phenomenology in his book on intuition: "what it is for an experience of yours to have presentational phenomenology with respect to *p* is for it to both make it seem to you that *p* and make it seem to you as if this experience makes you aware of a truth-maker for *p*" (2013, p. 37).

of a larger disease. The proposed phenomenological properties that supposedly explain why perceptual experiences can evidentially justify without being justified themselves are properties that are shared by experiences that are intuitively unable to provide justification. If my strong desire for gold influences my perceptual experience in a way that makes a yellow pebble look like gold, then this experience seems incapable of justifying my belief that I have found gold. Yet this token bad perceptual experience seems to have the same phenomenology as other good perceptual experiences.

Chudnoff does attempt to show that bad cases of cognitive penetration do not have the right kind of presentational phenomenology with respect to the relevant beliefs. According to Chudnoff (2013), a perceptual experience has presentational phenomenology *relative* to a certain proposition *p*, such that the experience makes it seem that *p* and makes it seem that the experience makes you aware of a truth-maker for *p*. The supposed solution for bad cases of cognitive penetration is that the experiences in these cases do not have presentational phenomenology *with respect to* the relevant propositions, because they do not seem to make you aware of the truth-makers for those propositions. For instance, in the case of an overly strong desire for gold, the perceptual experience simply does not seem to make you aware of a truth-maker for the proposition that you have found *gold* because it does not give you any awareness of the molecular structure in virtue of which the yellow pebble is gold (Chudnoff 2013, p. 91). And without presentational phenomenology with respect to a certain proposition, the experience alone cannot provide you with prima facie justification for that proposition.

Unfortunately, partly because this explanation does not have the broader Distinctiveness Problem in view, it falls short of providing a fully satisfying answer. First, if experiences do not seem to make you aware of truth-makers for propositions about gold, then there seem to be precious little propositions left for which experiences can make you aware of truth-makers. For instance, given that we are normally only aware of the facing surfaces of objects, one might claim that experiences do not make you aware of truth-makers for propositions about *objects*, but can only make you aware of truth-makers for propositions about facing surfaces. This would mean that dogmatism would lose much of its appeal. Second, the current solution would only work for cases in which cognitive penetration influences the high-level contents of perceptual experience. But there could also be cases of cognitive penetration in which low-level contents of experience, such as contents about colors, are influenced (see, e.g. Hansen et al. 2006). Chudnoff's explanation would do nothing to counter these cases of cognitive penetration, as experiences surely seem to make you aware of truth-makers for propositions about colors if they seem to make you aware of truth-makers at all.

The Distinctiveness Problem, and the associated problem of cognitive penetration, thus cannot be solved by an appeal to presentational phenomenology. In the next section we'll see that the phenomenological property of forcefulness faces a similar fate.

3.4.1.3 Forcefulness

The last phenomenalist answer to the Distinctiveness Problem is provided by Huemer, and appeals to the phenomenal property of *forcefulness*.[25] According to Huemer, perceptual experience not only has a certain representational content, it also presents this content in a certain way, namely *as being actualized*. Huemer explains what he means by forcefulness in more detail by contrasting it with what it is not:

> It is not a matter of how faint or vivid an experience is, and it is not a matter of how detailed and specific its content is. Forcefulness is not a matter of either the qualia or the representational content of an experience; it is a third aspect of experience. As to what that aspect is, I have no more to say than [the following]: it is the fact that, in the experience, it seems to one that something satisfying the content of the experience actually exists, here and now.
>
> (Huemer 2001, p. 79)

As with the earlier suggestions, we can reasonably ask whether the property of forcefulness is necessarily connected to perceptual experience, or whether other experiences could also have this phenomenal property.[26] Huemer suggests that it is when he claims that, because of the forcefulness of perceptual experience, "[...] you *would* [...] *never* confuse seeing a tomato with imagining one" (2001, p. 77; my italics). This is a very strong claim, as it surely seems possible that a subject somehow confuses seeing a tomato with imagining one. Moreover, that this possibility is also actual could be supported by appealing to the so-called 'Perky effect'.

Perky (1910) designed a set-up in which subjects were told to imagine a certain object, like a banana, book, or leaf, on a surface while a faint image which resembled that object was also projected (with increasing luminosity) onto the surface. Subjects did not know that an image was going to be projected, and also denied to have seen a projected image—they claimed they just imagined everything on the surface. Yet their reports made clear that the projected image did influence their experience: some noted that the banana was standing on end, which was not as they had been supposing they thought of it; some were surprised to imagine an elm leaf even though they had tried for a maple leaf (Perky 1910, p. 432). The precise interpretation of these findings remains controversial: subjects might have somehow mistaken their perceptual experiences (of the objects) for imaginations (of those objects), but the imagination could also have blocked the

[25]Chris Tucker (2010, p. 530) mentions a similar property of 'assertiveness', and William Tolhurst (1998, pp. 300–1) talks of the property of 'presence'.

[26]The problem of cognitive penetration could also be raised against this proposal again, as badly cognitively penetrated experiences are just as forceful as other perceptual experiences. However, Huemer responds to this problem by simply biting the bullet: according to him, cognitively penetrated experiences do provide immediate prima facie justification (Huemer 2013).

occurrence of a perceptual experience altogether (Segal and Gordon 1969). The former interpretation is clearly at odds with Huemer's claim that you would never confuse perception for imagination, and this would leave him without an explanation of how this is possible.

However, the latter interpretation, that imagination interfered to such an extent that no perceptual experience actually occurred, appears to be the usual one in current psychological literature.[27] Even if this is granted, then it is still surprising that a philosophical theory is a priori committed to a specific interpretation of the data. Moreover, the precise way in which imagination is supposed to interfere with perception remains unresolved (Reeves and Craver-Lemley 2012). Although there might not be a perceptual experience that is confused for an imaginatory experience, there might still be some confusion going on at the level of pre-phenomenal processing.[28]

Something similar could be said about confusing imaginations with perceptions. Some have claimed that hallucinations might just be internal mental events, like imaginations, that are mistaken for genuine perceptions because of a diminished skill in 'reality discrimination' (Johnson et al. 1993; Fish 2008). Again the idea does not seem to be that subjects consciously assess their experience and wrongly judge them to be either perception or imagination, but that there is some unconscious mechanism that somehow flags them as the one or the other.[29]

It is difficult to assess the precise philosophical force of these empirical hypotheses, but a few things are clear. First, subjects in the Perky experiment lack forceful experiences, while hallucinating subjects have forceful experiences. Huemer would probably accept these consequences as well, concluding that hallucinating subjects have justification for their beliefs (which Huemer explicitly endorses and takes to be a virtue of his theory (2001, p. 129)), while subjects in the Perky experiment lack such justification. Second, the presence or absence of forcefulness in these examples should not be reducible to a higher-order belief on the part of the subject with the content that, e.g., the current experience is perceptual. If it were reducible to such a higher-order belief, then dogmatism couldn't gain any epistemic mileage out of it: this belief would have to be justified itself before it could evidentially justify any other beliefs, and even if it was justified it would conflict with the supposed *immediacy* of perceptual justification.

Huemer would certainly agree with this second point, as he holds that forcefulness is a specific type of seeming (i.e., a seeming that something satisfying the

---

[27]Although this is not so easy to determine, as the 'Perky effect' nowadays refers in psychology to the general interference of imagination on perception. Although it is implausible to explain this *general* effect by appealing to confusion of perception with imagination, in specific instances (viz., the types of experiences in the 1910 Perky experiment) this confusion might still take place.

[28]Cf. Reeves and Craver-Lemley (2012, p. 7): "An [attractive] alternative hypothesis is that the Perky effect results from a combination of real and imagined features that makes the real features more difficult to extract."

[29]I discuss the possibility of such an introspective mechanism in more detail in Chap. 6.

content of the experience actually exists) and that in general "its seeming to *S* as if *P* is a distinct state from *S*'s believing that *P*" (Huemer 2001, p. 99). As it appears implausible to posit a non-belief-like intellectual seeming that something satisfying the content of experience actually exists in addition to the perceptual seeming that such-and-so is the case, the best way to interpret Huemer is to take him as suggesting that there is a phenomenal property of experience, not reducible to the content of the experience, which conveys that external objects are actually present. But the problem is that we have been given no explanation at all how this irreducible phenomenal property of experience (dis)appears in the empirical cases under discussion. So even in the best scenario, dogmatism is left with an explanatory challenge.

This is especially pressing because one can give an ontologically simpler explanation of the relevant phenomena. In the Perky experiments, subjects do not have any forceful experiences because they implicitly believe that they are imagining instead of perceiving. In the case of hallucination, subjects implicitly believe that they are perceiving whatever it is that they might instead be imagining, and therefore have forceful experiences. This is an explanation that appeals to higher-order beliefs, instead of supposing the presence or absence of an extra irreducible phenomenal property to the content of experience. And if forcefulness indeed comes down to having certain higher-order beliefs, then it is of no help to dogmatism.

In fact, the same type of explanation can be provided for all of the proposed distinctively perceptual phenomenologies. Why do perceptual experiences, in contrast with imaginations, usually have an associated phenomenology of seeming to ascertain that *p*, or seeming to make you aware of a truth-maker for *p*? Because subjects usually know that they are perceiving or imagining, and on that basis believe that they either are, or are not, taking in the environment as it is. If the proposed distinctive phenomenology of perceptual experience comes down to higher-order beliefs of this sort, then dogmatism cannot explain how experience is able to provide immediate prima facie justification.

Note that explaining typical perceptual phenomenology in terms of higher-order beliefs raises a question about the origins and grounds of these higher-order beliefs. If a higher-order belief that I am currently perceiving that *p* explains the phenomenology that is usually associated with perception, then this phenomenology is apparently not used as a *basis for* those higher-order beliefs. I think that's exactly right, but we will return to this issue in Chap. 6.

3.4.2 General Worries About Phenomenalism

From the specific discussions of the several phenomenalist answers to the Distinctiveness Problem we can distill a general recipe that phenomenalists have to follow to come up with a satisfying reply. First, the phenomenalist has to describe a phenomenal feature that is distinctive of those perceptual experiences that are intuitively able to provide immediate justification. This feature should not be shared with either imaginations (incapable of evidential justification), beliefs (only capable of evidential justification if justified themselves), or what we may call ‘bad’ perceptual experiences (such as those in bad cases of cognitive penetration). None of the proposed phenomenal properties comes even close to fulfilling these desiderata, and most probably because dogmatists have not realized the full extent of the Distinctiveness Problem.

Second, the phenomenalist has to show that his or her preferred phenomenal feature is not reducible to a (higher-order) belief. If the feature is reducible to such a belief, then this belief will either be itself unjustified and therefore unable to evidentially justify the perceptual belief, or justified, but then the justification of perceptual beliefs is no longer immediate. I think that many of the phenomenal features proposed to be distinctive of perceptual experience can actually be reduced to higher-order beliefs that one is currently perceiving. On this view, the distinctive phenomenology of perception is thus not the basis for but rather the result of a higher-order identification process.

Third, even if one is able to provide an irreducibly distinctive phenomenal feature of those experiences that are capable of providing immediate justification, then the dogmatist is not home free yet. Take whatever phenomenal feature you like, be it a feeling of seeming to ascertain that *p*, or presentational phenomenology, or forcefulness, the phenomenalist will still have to show that this feature is also *epistemically* significant. After all, the Distinctiveness Problem does not just ask for any specific feature that distinguishes good perceptual experiences from beliefs, desires and badly cognitively penetrated experiences. What is needed to answer the problem is a feature that is both specific to good perceptual experiences *and* epistemically significant. And this remains problematic. Even if it phenomenally seems to you that you are aware of all the items that make the currently perceptually represented proposition true, why should this provide you with justification for believing that proposition? How could phenomenology somehow add justificatory strength to the represented propositions?

Here the proponent of dogmatism might claim that explanations have to stop somewhere and that the envisaged justificatory relation between contentful phenomenal experience and perceptual belief is primitive.[30] However, in the face of relevant externalist alternatives, such a reply might not do the trick. The externalist *can* supply an explanation of why, for instance, the reliability of a belief-forming process is important for justification. Reliability is important because it makes sure that justified beliefs have some connection to *being true*, and that is an important epistemic value. Such a further explanation of the epistemic relevance of phenomenology will also be important to provide if one wants to defend the view that the phenomenology of experience provides it with a distinctive justificatory power.

[30]See for instance (Siegel 2013b, p. 757) for this type of reply.

Chudnoff (2013) has risen to this challenge, and maintains that having presentational phenomenology is epistemically significant because of its relation to being in a position to know. According to Chudnoff, if one's perceptual experience puts one in a position to know that *p*, then it does so because its presentational phenomenology is veridical (Chudnoff 2013, p. 174). It is in virtue of the fact that *p* and the fact that one's experience makes one aware of a truth-maker for *p* that one is in a position to know that *p*.[31] Given that all experiences with presentational phenomenology thus instantiate a type of phenomenology that is crucial to being in a position to know, subjects of these experiences are still reasonable to take them at face value even if the experiences are not veridical. After all, from the inside it is to these subjects as if they really are in a position to know.

However, even if we grant the point about being in a position to know in virtue of veridical presentational phenomenology, then the proposed explanation of the justificatory power of perceptual experience still falls short. And this is because the kind of reasonableness that it appeals to is not sufficient for epistemic justification. Although it seems unreasonable not take an experience at face value when the experience has presentational phenomenology and one has no reason to distrust it, this is simply not enough to acquire a justified belief on the basis of the experience. Once one has acquired an unjustified belief, one would also be unreasonable not to take this belief into consideration when forming new beliefs in the absence of any reasons to distrust the unjustified belief. Nevertheless, the new beliefs will not be justified if they are based on the earlier acquired unjustified belief. The type of reasonableness in question should thus not be equated with justification.[32]

Given that none of the above three steps for solving the Distinctiveness Problem has been adequately carried out by dogmatism, it remains in full force. But that means that dogmatism is, at least for the moment, a problematic theory of perceptual justification. Just like evidentialism, dogmatism appears unable to overcome the problems presented by the Sellarsian dilemma.

3.5 The Speckled Hen Revisited

In the previous section I argued that none of the phenomenalist answers to the Distinctiveness Problem were convincing. First, the phenomenal properties that were supposed to explain the distinctive justificatory power of experience were not distinctive of experiences that are intuitively capable of providing immediate evidential justification. Second, the proposed phenomenal properties might be

[31] Note that Chudnoff also discusses some apparent counter-examples to this claim, such as environmental Gettier cases.

[32] McGrath (2013a) makes this point against proponents of dogmatism who respond to bad cases of cognitive penetration by pointing out that the subjects are still entirely reasonable to believe what they do.

reducible to higher-order beliefs of a certain sort, in which case the immediacy of perceptual justification is challenged. Third, the epistemic significance of the proposed phenomenal properties remains unclear. This section focuses on an example to bring out more clearly *why* dogmatism's sufficiency claim is doomed to fail without a reliabilist component.

The example to be discussed is a variant of the speckled hen problem. Although I argued in Sect. 3.3 that dogmatists can provide a plausible story about the difference between seeing a 3-speckled hen and seeing a 48 speckled hen on the basis of differences in the representational content of the experiences, the speckled hen can still be used to bring out another problem for dogmatism (and there is some indication that Sosa also intended to raise this problem with the example).

Suppose a subject looks at a speckled hen with 48 speckles, has a perceptual experience representing that the hen has *around* 50 speckles, and comes to believe that the hen has around 50 speckles. Then, according to dogmatism, the belief of this subject is immediately justified on the basis of the experience. However, now suppose that the subject is actually generally inept in making such judgments (cf. BonJour and Sosa 2003, p. 127, n. 7): in many cases where he has an experience which represents that the hen has around 50 speckles, the subject comes to hold false beliefs about the number of speckles of the hen (e.g., that the hen has precisely 50 speckles, or that the hen has around 30 speckles). The question that arises is this: given that the subject is generally inept in forming beliefs on the basis of his perceptual experience, why should we think that he is justified in his belief the one time he gets it right?

A first response to this specific case might claim that dogmatism is primarily a theory about propositional rather than doxastic justification. That is, although in these cases the subject's beliefs have propositional justification because of the evidence constituted by the relevant perceptual experiences, the beliefs are not doxastically justified because they are not properly *based* on the experiences. This is the type of response that is favored by Smithies (2012a) and Chudnoff (2013). On this line of thought one can have justifying evidence to believe that p without having the capacity to actually base a belief on this evidence. What one needs for such a capacity presumably is a *reliable* disposition to go from the evidence that p to a belief that p.

Although it's fairly plausible to hold that one lacks doxastic justification if one cannot reliably form the belief that p on the basis of evidence that p, it's not so clear that in such a case it would still make sense to ascribe propositional justification. In what sense does a subject really *have* or *possess* the evidence that p if he has no way of basing beliefs on this evidence? What's more, this kind of reply does not help with a related problem where matters are approached from the stimulus-side of experience.

The previous problem had to do with the fact that there might not be a reliable connection between the perceptual experience that p and the belief that p. Now we look at an example where there is no reliable connection between the fact that p and the perceptual experience that p. Suppose that there is a subject, Sherri, who has a perceptual experience that there is a cherry before her whenever she either looks at a cherry or looks at a grape.[33] As it so happens, Sherri is looking at a grape, her perceptual experience represents that there is a cherry before her, and on that basis Sherri believes that there is a cherry before her. According to dogmatism, Sherri would be immediately justified in her belief, as she has an experience that there is a cherry before her, no reason to distrust this experience, and has properly based her belief on the experience. However, given what we know about the malfunctioning of her perceptual system, it seems as though Sherri's belief is not really justified either propositionally or doxastically. One might even go further and hold that the experience's lack of sensitivity to the environment is sufficient for undermining justification even in situations where Sherri is actually looking at cherries instead of grapes.

The intuitive pull to take Sherri's beliefs to be unjustified can be explained by the lack of any reliable connection between her experience that there is a cherry before her and the fact that there actually is a cherry before her. If we combine this with the insight that there should be a reliable connection between the experience that there is a cherry before her and the belief that there is a cherry before her, we can come to the tentative conclusion that what is ultimately needed is some sort of a reliable connection between the *fact* that there is a cherry before her and the *belief* that there is a cherry before her. Contrary to what dogmatism maintains, epistemic justification requires more than just having an experience with the right content.

However, perhaps more can be said on behalf of dogmatism. Perhaps the envisaged situations in which contentful experiences are not properly connected to the facts are impossible because of the nature of perceptual content. On this line of thought, it's simply impossible that a perceptual experience has the content that *p* if it's not reliably caused by the fact that *p*. If, say, both a cherry and a grape cause the same perceptual experience, then this experience cannot have a content that specifically represents cherries. It will at best represent 'chapes'. Similarly, perhaps it's impossible that a perceptual experience has the content that *p* but does not reliably lead to the belief that *p* under normal circumstances (in which, e.g., there is no countervailing evidence and there are no cognitive malfunctionings).

Now it would really take us too far afield to argue for a specific view of representational content, so I will not consider the details of this proposal.[34] The important point is that this proposal will make some large concessions on behalf of dogmatism towards externalism.[35] Once one elucidates dogmatism by acknowledging that a perceptual experience that *p* must be reliably connected to both the fact that *p* and the belief that *p*, one acknowledges in fact that reliability is also needed to obtain a sufficient condition for justification. But what was so

[33] If you find the example hard to imagine, you could treat it as a bad case of cognitive penetration where Sherri's strong desire for cherries makes grapes look like cherries. But there could also be a different kind of cognitive malfunctioning that simply ends in Sherri's having cherry-experiences whenever she sees grapes.

[34] But see e.g., Millikan (1989) for objections.

[35] The same can be said for Brogaard's (2013b) sensible dogmatism.

interesting about dogmatism was that it provided a sufficiency claim for justification that, on the face of it, had nothing to do with reliability, hence, that it appeared to claim that reliability was *not necessary* for perceptual justification. That is also why so many of its proponents look at the *phenomenology* of perceptual experience to explain where it gets its distinctive justificatory power from.

Nevertheless, it is of course possible to go for this kind of reliabilist dogmatism. This would make the built-in reliability of a perceptual experience that *p* in combination with the conscious experience that *p* sufficient for immediate prima facie justification of the belief that *p*. Note, though, that such an account loses the motivation from the New Evil Demon Intuition. According to this intuition, subjects who are radically deceived by an evil demon or malevolent scientist still have as much justification for their beliefs as we do for ours as long as they have the same experiences as us. The current reliabilist version of dogmatism would not just have to claim that this intuition is mistaken, it would have to say that the entire scenario is *impossible*: given that the experiences of the deceived subjects are not reliably connected to facts in the world, they cannot have the same content as our experiences.

Another problem for reliabilist dogmatism in general is that it is no longer clear that experience is doing any epistemic work. Perceptual experience with built-in reliability is supposed to be sufficient for justification, and we've seen that there are problems with views that make a sufficiency claim about justification without the built-in reliability of perceptual experience. So why not suppose that, in fact, reliability is all that matters for justification, and experience is an unnecessary component?

The only thing left to motivate the necessity of experience would be its explanation of the Blindsight Intuition, i.e., that a blindsighted subject is not justified in his beliefs in the absence of perceptual experience. But notice that, at this point, the dogmatist will have to take the necessity of experience as a *brute* fact about perceptual justification. Without an answer to the Distinctiveness Problem, the dogmatist simply accepts the Blindsight Intuition at face value without providing any theoretical underpinnings. This would thus make the motivation for reliabilist dogmatism fairly weak.

One might still try to salvage reliabilist dogmatism by appealing to the reliability of experience as the property that answers the Distinctiveness Problem. According to this explanation, experience is able to provide justification for belief precisely because its content is reliably connected to how the world is. But this leads back to the previous worry: is experience really doing any epistemic work, or is it only reliability that matters? This becomes more clear once one rephrases the current suggestion in the following way: experiences can evidentially justify beliefs when they are themselves justified, and their own justification arises because of their being reliably connected to the world. If this is the suggestion, then it is reasonable to ask why reliability by itself would be sufficient to justify experience, but not belief. Again, reliabilist dogmatism seems rather unmotivated as a general theory of perceptual justification without a fuller explanation of the necessity of experience for perceptual justification.

3.6 A Final Argument for Dogmatism?

So far I have presented several severe challenges for dogmatism. The Distinctiveness Problem shows that dogmatism needs to do a lot more work if it is to overcome the objection that it is an ad hoc theory of perceptual justification. The new problem of the speckled hen shows that dogmatism will have counter-intuitive consequences without incorporating some sort of reliabilist criterion. However, one could respond that these are mere challenges for dogmatism if one also has a good argument to show that dogmatism is actually the only theory of justification that allows us to have justified beliefs. In that case, dogmatism would be the only game in town.

Huemer has presented an argument that tries to show exactly that (Huemer 2007). The argument can be briefly put as follows:

1. When we form beliefs, with a few exceptions not relevant here (i.e., self-deception, brain-damage, insanity), our beliefs are based on the way things seem to us (Huemer 2007, p. 39).
2. If one's belief that *p* is based on something that does not constitute a source of justification for believing that *p*, then one's belief that *p* is unjustified (Huemer 2007, p. 40).

C. If [...] appearances do not confer at least some defeasible justification on propositions that are their contents, then our beliefs are generally unjustified (Huemer 2007, p. 41).

Note that I will only consider this argument as far as perceptual justification is concerned, even though Huemer tries to use it for justification in general. That means that I will assume that we are talking about perceptual beliefs, and that "the way things seem to us" should be interpreted as referring to perceptual experiences with specific contents.

If Huemer's argument is correct, then some version of dogmatism (although it could be a reliabilist version) must be true if our perceptual beliefs are to be justified. If we base our perceptual beliefs on our experiences, but experiences are unable to confer justification on these beliefs, then all of our perceptual beliefs would be unjustified.

For the sake of clarity, I'll immediately state where I think the argument goes wrong: the argument suffers from a fallacy of ambiguity. Premise 1 is only plausible if the basing relation is interpreted as a mere causal relation, but premise 2 needs something stronger than a mere causal relation to be true. To show this, I'll begin reviewing the argument with the mere causal interpretation of the basing relation in mind, and then go on to discuss the stronger interpretation.

Let's start with reviewing the first premise of Huemer's argument. According to Huemer, this premise should be understood as an *empirical* thesis about what we generally base our beliefs on. Although this looks plausible if we interpret the basing relation as a mere causal relation, there is still reason to contest it. As Michael DePaul (2009) puts it, this thesis is a piece of "armchair psychology" that might very well turn out to be false. For instance, perceptual experience and perceptual belief might both stem from a different common cause, or they might even have to do with completely distinct cognitive processes.[36] Introspection will not be able to decide between these competing hypotheses. However, for the sake of argument, let's grant that perceptual experience is indeed the proximate cause of perceptual belief.

We can now go on to premise 2. Given that we are interpreting the basing relation as a mere causal relation, premise 2 is obviously false. The firing of some set of neurons is probably responsible for causing perceptual beliefs without the firing of these neurons being a source of justification for these perceptual beliefs. Surely this does not preclude these perceptual beliefs from being justified. So if we interpret the basing relation as a mere causal relation, premise 1 is contestable but at least looks plausible, while premise 2 is obviously false.

Premise 2 is far more plausible if we interpret the basing relation as something stronger than a mere causal relation. If I believe that *q* on the basis of the belief that *p* and the belief that if *p*, then *q*, then my belief will not be justified if the beliefs on which I base it are not justified, i.e., if those beliefs do not together constitute a source of justification.

But if we now return to premise 1 with this stronger interpretation of the basing relation in hand, then the premise is doubtful for several reasons. First, notice that the earlier problem about the causal interpretation of premise 1, that perceptual beliefs might not be caused by perceptual experiences, remains problematic for the stronger interpretation. Even if a causal relation is not sufficient for the basing relation, it surely is necessary.

Second, Huemer provides little reason to support premise 1 beyond the causal interpretation. Huemer merely argues that perceptual experiences are normally "the only (proximately) causally relevant factor in one's belief-formation" (Huemer 2007, p. 39). But even if this is true, this does not show that the perceptual experiences are thereby the only factor on which our beliefs are based. After all, distal causes are still seen as causes for beliefs, so why couldn't these distal causes also be part of the justificatory basis for these beliefs?

Moreover, it could be the case that perceptual beliefs have *no* basis. Recall the distinction between evidential and non-evidential justifiers. Only evidential justifiers are bases for beliefs, and one could adopt a view according to which perceptual beliefs have no evidential justifiers. Indeed, this is precisely what reliabilists like Lyons (2009) might maintain: beliefs are reliably caused by perceptual experiences but are not evidentially justified by them, i.e., are not based upon them. Huemer apparently assumes that perceptual beliefs must be based on something, which makes it natural to equate the relevant proximate causes with bases. That is also why he is so focused upon showing that, e.g., "the actual reliability of a process has no independent causal relevance" (Huemer 2007, p. 48). But whether perceptual beliefs have bases at all is precisely one of the things at issue between dogmatists and externalists, and so cannot be assumed without begging the question.

[36]See Lyons (2009, pp. 60–1) for more on this line of thought.

In fact, the externalist idea of baseless perceptual beliefs is phenomenologically better supported than the dogmatist idea of experience-based beliefs. In general, we are simply not aware of basing our perceptual beliefs on our perceptual experiences. This does not just amount to the claim that we do not consciously reason from "it seems to me that *p*" to "*p*" (recall that Pryor explicitly agrees with this—Sect. 3.2). It comes down to the phenomenological fact that we are usually not aware of two separate mental events, viz. having a perceptual experience and forming a belief, let alone that we are aware of basing these beliefs on our experiences. Normally, perceiving just *is* believing. We only become aware of a difference between experience and belief when we are aware of some sort of defeater, e.g., when we are familiar with the illusory nature of a certain experience as in the case of the Müller-Lyer illusion. This is no knock-down argument against the thesis that we do base our beliefs on our experiences, but it does show that this is a substantive philosophical thesis instead of a strong phenomenological fact.

Premise 1 is thus in need of far more argumentative support before it can be accepted as a thesis about a strong basing relation between perceptual experiences and perceptual beliefs. But it certainly looks more plausible if one has a weaker, causal relation in mind. Premise 2, on the other hand, is only plausible when interpreted as being about a strong basing relation, and is clearly false if interpreted in the causal sense. The argument thus fails in its entirety even though each of its premises can be given a reading that provides it with some intuitive support.

3.7 Conclusion

According to dogmatism, having a perceptual experience that *p* is sufficient for immediate prima facie justification of the belief that *p*. By committing itself to the propositional content of experiences, dogmatism thus grasps the second horn of the Sellarsian dilemma outlined in the previous chapter. Unfortunately, it is also unable to overcome the problems associated with this horn of the dilemma. Although dogmatism can appeal to a difference in representational content to explain the epistemic difference between novices and experts and between seeing a 3-speckled and a 48-speckled hen, it cannot explain the posited epistemic difference between experience and belief. Specifically, dogmatism does not have an answer to the Distinctiveness Problem: it cannot explain what is so distinctive of epistemically good perceptual experiences that allows them to evidentially justify beliefs (in contrast with e.g., imaginations, desires, and badly cognitively penetrated experiences) without being justified itself (in contrast with belief).

Dogmatists have been quick to point to several phenomenal properties of perceptual experience that supposedly answer the Distinctiveness Problem. These proposals all fail because the phenomenal property is either not distinctive of perceptual experiences that are intuitively able to provide immediate justification, or else reducible to or possibly caused by higher-order beliefs. Moreover, the dogmatist is in a dialectically weak position because the epistemic significance of phenomenology is left unexplained.

That phenomenology is not sufficient for justification in the absence of reliability comes out clearly in the new problem of the speckled hen: a perceptual experience that *p* might not be reliably connected to either the fact that *p* or the belief that *p*. In both cases the belief that *p* appears intuitively unjustified even if it is based correctly on the experience that *p*. A reliabilist dogmatism might solve this problem, but this would (a) lose the motivation from the New Evil Demon Intuition, and (b) raise the question of whether reliability could be sufficient for justification all by itself.

The above problems might merely be challenging for dogmatism if it was the only viable account of perceptual justification. Huemer's argument, designed to show just that, fails to make good on its promise because it suffers from a fallacy of ambiguity: perceptual experiences might cause perceptual beliefs, but they need not be *based* on them in any stronger sense of the term. This opens the floor to competing accounts of perceptual justification.

The first of these competing accounts is presented in the next chapter: *epistemological disjunctivism*. It agrees with experientialism that perceptual beliefs are evidentially justified, but it disagrees about the nature of the justifier. Instead of holding that a perceptual experience is itself the reason in virtue of which one has justification for one's beliefs, this view holds that it is the fact that one sees that *p* in virtue of which one has justification for believing that *p*. However, in order to use this fact as a reason, epistemological disjunctivism brings with it a requirement for *higher-order* beliefs. Although this requirement leads to problems in the end, it will bring us one step closer to the externalist account of perceptual justification I aim to defend.

Chapter 4
Epistemological Disjunctivism and Higher-Order Issues

4.1 Introduction

In the previous two chapters I argued against variants of experientialism, which hold that experiences evidentially justify perceptual beliefs. In this chapter we look at two variants of an alternative view of both perceptual knowledge and justification, called epistemological disjunctivism. Now, on some presentations of epistemological disjunctivism, the view merely holds that one's experiential evidence is different between good and bad cases of perception. That is, one's experiences evidentially justify beliefs, but the experiences one has in successful cases of perception are better evidence than the experiences one has in indistinguishable unsuccessful cases of perception. I will not focus on this version of epistemological disjunctivism, as it encounters many of the objections of the previous chapters—for instance, how can experiences constitute evidence at all without being justified, and what explains why experiences in successful cases of perception are better evidence than experiences in bad cases of perception? Instead, the variants of epistemological disjunctivism that will be the focus of this chapter hold that perceptual beliefs are justified by factive reasons of the form "I see that *p*". These accounts thus retain the idea that perceptual beliefs are evidentially justified, but change the evidential justifiers from experiences to factive reasons.

In Sect. 4.2, I discuss Duncan Pritchard's (2012b) presentation of epistemological disjunctivism and some of its purported motivations. On Pritchard's view, the fact that *S* sees that *p* provides reflectively accessible rational support in virtue of which *S* knows that *p*. Pritchard can thus be taken to uphold a Justified True Belief version of Epistemological Disjunctivism (JTBED), even though the proposed justifier is different from traditional JTB accounts. This also portrays one of Pritchard's main motivations for the view: JTBED supposedly captures key elements of both internalism and externalism by combining the accessibility of reasons with their factivity.

H. Ghijsen, *The Puzzle of Perceptual Justification*, Synthese Library 377,
DOI 10.1007/978-3-319-30500-4_4

In Sect. 4.3, I raise three problems for JTBED. The first problem is one that Pritchard introduces himself: the basis problem. If seeing that *p* constitutes the reason, or basis, in virtue of which one knows that *p*, then seeing that *p* can no longer be analyzed as being a specific way of knowing that *p*. The second problem has to do with Pritchard's contention that JTBED captures the key elements of internalism and externalism. I will argue against this contention by showing that JTBED fails to accommodate the New Evil Demon Intuition (a key element of internalism), and also fails to provide a good account of animal knowledge (a key element of externalism). The third problem for JTBED is that it simply lacks a satisfactory account of how we access factive reasons.

In Sect. 4.4, I present an alternative version of epistemological disjunctivism proposed by Alan Millar (2010, 2011). The main difference between this version and that of Pritchard is that it takes knowledge to be the primitive notion in terms of which justification can be explained. Millar thus proposes a Knowledge First version of Epistemological Disjunctivism (KFED) rather than a JTBED. According to Millar's KFED, seeing that *p* does not constitute the kind of rational support *in virtue of which* one knows that *p*, instead, it only constitutes a factive reason that can be used to justify one's belief that *p*.

In Sect. 4.5, I show how KFED can avoid most of the problems that plagued JTBED. However, KFED faces some problems of its own. While intuitively knowledge is logically stronger than justified belief, by Millar's lights, it turns out to be weaker: knowledge does not entail justified belief, but justified belief does entail knowledge. Even though both versions of epistemological disjunctivism are thus ultimately unsuccessful, they do have some important insights regarding the importance of factive reasons. Section 4.6 discusses these insights and relates them to a possible way of explaining the Blindsight Intuition.

The final Sect. 4.7 sums up the results of this chapter and concludes from this that perceptual beliefs might not be *evidentially* justified at all.

4.2 Pritchard's JTBED

The core thesis of Pritchard's epistemological disjunctivism is as follows:

> In paradigmatic cases of perceptual knowledge an agent, *S*, has perceptual knowledge that ϕ in virtue of being in possession of rational support, *R*, for her belief that ϕ which is both *factive* (i.e., *R*'s obtaining entails ϕ) and *reflectively accessible* to *S*.
>
> (Pritchard 2012b, p. 13)

According to Pritchard, an agent *S* has paradigmatic perceptual knowledge *in virtue of* having a specific type of justification which ensures that the perceptual beliefs it supports are true.[1] Perceptual knowledge is thus analyzed as justified, true

[1] In what follows, I will take Pritchard's restriction to paradigmatic cases of perceptual knowledge as read.

belief, although the type of justification is different from the type of justification that traditional JTB accounts have appealed to. Traditional accounts appealed to *fallible* reasons to provide the justification required for knowledge, reasons that are in principle compatible with the falsity of the belief in question—although knowledge of course requires that the belief in question is in fact true. In contrast, according to JTBED the relevant type of justification has to do with having reflectively accessible *factive* reasons, that is, reasons which entail that the beliefs they support are true.

The combination of factivity and accessibility is what makes JTBED counterintuitive. According to JTBED, a subject has reflective access to a factive reason in an epistemically good case, while it merely seems to the subject as if he had reflective access to such a reason in an introspectively indistinguishable epistemically bad case (e.g. in an hallucination). Pritchard goes to great lengths to argue that factivity and accessibility can be combined, but has surprisingly little to say about what exactly it means to have reflective access other than that "the subject can come to know through reflection alone that she is in possession of this rational support" (ibid.). I will return to this point later on (Sect. 4.3.3).

Let's get a little more concrete about the kind of reasons in play in perceptual knowledge. According to Pritchard:

> The particular kind of rational support that the epistemological disjunctivist claims that our beliefs enjoy in paradigm cases of perceptual knowledge is that provided by *seeing that* the target proposition obtains. So when one has paradigmatic perceptual knowledge of a proposition, *p*, one's reflectively accessible rational support for believing that *p* is that one *sees that p*. Seeing that *p* is factive, however, in that if it is the case that one sees that *p* then *p* must be true.
>
> (Pritchard 2012b, p. 14)

According to Pritchard, seeing that *p* is, at least for the case of visual perception, what supports perceptual knowledge of *p*. For instance, when a subject *S* is looking at a tree in epistemically favorable conditions such that there are no non-tree look-alikes in the environment, *S*'s perceptual faculties are working correctly, no deceivers are present, etc., then *S*'s rational support for the belief that there is a tree is the fact that *S* sees that there is a tree. This fact is reflectively accessible to *S* and entails that there is a tree. In contrast, when a subject S' hallucinates a tree, then S' does not see that there is a tree, so the same kind of rational support is not present to be accessed by S'. Of course, S' might think that he has the same kind of rational support while hallucinating a tree, but thinking that one has a certain reason comes apart from actually having this reason on the epistemological disjunctivist's picture.

What motivations are there to uphold such a view of perceptual knowledge? A first motivation is that it can easily accommodate our ordinary way of talking and thinking about perceptual knowledge. If someone challenges your claim to know that, e.g., your mutual friend is also at the party on the grounds that he usually does not go to parties, then you might well support your knowledge claim by saying that you know your friend is here *because* you see that he is standing over there. In this scenario, you justify your claim to know that your friend is at the party precisely by appealing to the factive reason that supported your knowledge in the first place. At least, that is a natural way to interpret what is going on in the imagined case.[2]

The second, and, according to Pritchard, most important motivation for JTBED is that it can capture key elements of both internalism and externalism in epistemology. It is for that reason that Pritchard portrays the view as being "the holy grail of epistemology" (2012b, p. 1). Perceptual knowledge has to do with having reasons that are reflectively accessible, just as the internalist wants, but those reasons are also factive, thus securing the connection between epistemic support and truth that has been stressed by externalists. The first element allows for taking epistemic responsibility for one's beliefs, while the second makes sure that one cannot be justified in a body of beliefs that is still entirely unconnected to the truth.

This motivation is especially important because many views are unable to accommodate both elements, thereby giving the impression that one should pick only one of them. Take dogmatism as an example of a theory that only incorporates the internalist element. According to dogmatism, having an experience that *p* gives one immediate prima facie justification for the belief that *p*.[3] In this theory, a subject's epistemic support for the perceptual belief that *p* (i.e., his experience that *p*) is certainly accessible, but it does not guarantee that *p* is actually true: one can have an experience that *p* without *p* actually being true.

On the other hand, one might take a simple form of reliabilism as a theory that only incorporates the externalist element (Goldman 1979).[4] According to this simple reliabilism, a belief is justified if and only if it is the output of a reliable cognitive process, where a reliable cognitive process is a process that tends to lead to true beliefs. Although this guarantees that there is a close connection between a belief's being justified and a belief's being true (although not one of entailment), it also has the consequence that a subject can have a justified belief without knowing it. This is so because the subject might not have any access to the reliability of the cognitive process that leads to his belief.

If JTBED can, in contrast with these previous theories, indeed incorporate all key internalist and externalist elements, then it would surely be a strong contender as a theory of perceptual knowledge and justification. However, Pritchard's epistemological disjunctivism encounters a number of serious problems that ultimately make it unsatisfactory. This will be the topic for our next section.

---

[2]Note, however, that even Pritchard acknowledges that it might not be the only way (2012b, pp. 17–8).

[3]See Chap. 3.

[4]I will discuss reliabilism in more detail in the next chapter.

4.3 Three Problems for JTBED

Although Pritchard portrays JTBED as the holy grail of epistemology, in this section I will list three problems that point to a rather different conclusion. First, JTBED cannot adequately deal with the basis problem, second, JTBED does not incorporate several key internalist and externalist elements, and third, JTBED cannot provide a good account of access to factive reasons.

4.3.1 The Basis Problem

Recall that Pritchard claims that the fact that one sees that *p* is the rational support *in virtue of which* one knows that *p* in paradigm cases of perceptual knowledge. But this creates a difficulty for his view if we combine it with the following popular and intuitive conception of seeing that *p*:

SK Seeing that *p* is a way of **k**nowing that *p* (e.g., Williamson 2000; Cassam 2007; Millar 2010).

If SK is correct, then seeing that *p* cannot be the rational basis in virtue of which one knows that *p*. This would make the perceptual knowledge that *p* literally self-supporting, which appears to be an unacceptable view of empirical knowledge. This is the problem that Pritchard dubs as the *basis problem*.

The obvious way out of this problem is to drive a wedge between seeing that *p* and knowing that *p*. In that case, the former can be the factive reason that supports the latter without already constituting the latter. However, the wedge between seeing and knowing cannot become too great, because Pritchard also wants to accommodate an important epistemological intuition about the following case.

Fake Barn County: Barney is driving through a county in which most structures that look like barns are in fact mere façades, albeit ones that are constructed so cleverly as to be indistinguishable from real barns from Barney's position on the road. Barney happens to look at a real barn from his position on the road, and, not knowing that he is in fact in fake barn county, forms the belief that the object in front of him is a barn.[5]

Many philosophers have the intuition that in this scenario, Barney does not know that the object he is looking at is a barn. He would have believed the same thing had he been looking at a fake barn, and that situation could have easily occurred because there are so many fake barns around. Barney is simply unfortunate to be in epistemically bad circumstances that preclude him from knowing that the object in front of him is a barn.

Pritchard wants to accommodate this intuition, and so his maneuvering space in response to the basis problem is limited. If Barney is able to see that the object

[5]The example is from Goldman (1976).

in front of him is a barn, then Barney will presumably have access to a factive reason that allows him to know that the object in front of him is a barn. Given that Pritchard seeks to accommodate the intuition that Barney does not know that the object in front of him is a barn, Pritchard has to provide an alternative to SK on which Barney does not see that the object in front of him is a barn. This means that accounts that reduce seeing that *p* to veridically representing that *p*, or standing in the right causal connections to the fact that *p*, are all out of the question: on those accounts, Barney would see and therefore know that *p*.

The challenge for Pritchard thus is to present an account of seeing that is less epistemically demanding than SK, but sufficiently epistemically demanding to prevent Barney from seeing that *p*. Pritchard proposes the following:

SPK **S**eeing that *p* is (a specific way of) being in a good **p**osition to **k**now that *p* (Pritchard 2012b, p. 26).[6]

Although SPK gives JTBED an account of seeing that *p* that avoids the basis problem, it's constructed in a way that makes it feel ad hoc. Pritchard realizes that this is the case, and so attempts to provide it with some additional support. Specifically, Pritchard contends that the following case is an intuitive counterexample to SK that can nevertheless be accommodated by SPK:

> Suppose [. . .] that one is in a situation in which one is genuinely visually presented with a barn and circumstances are in fact epistemically good [. . .]. But now suppose further that one has been told, by an otherwise reliable informant, that one is presently being deceived even though this is in fact not the case [. . .]. Clearly, in such a case one ought not to believe the target proposition [that there is a barn before one], and hence one cannot possibly know this proposition either.
>
> (Pritchard 2012b, p. 26)

The subject of such a scenario, call him Barnaby, plausibly does not know that there is a barn before him as he does not, or at least should not, believe that there is a barn before him. Yet Pritchard maintains that he does *see* that there is a barn in front of him, thereby making it into a counterexample to SK. Pritchard's reason for this claim is that, upon discovery that the received testimony was false, Barnaby *would treat himself as having earlier seen that* there was a barn before him (ibid). According to Pritchard, this is a very natural way for him to describe his own situation once he has learned of all the relevant facts.

If the example works, then Pritchard has presented a challenge to SK. To serve as a separate motivation for SPK though, one also has to show that SPK, in contrast

[6]Note that Pritchard puts it specifically as "being in a state that guarantees that one is in a good position to gain knowledge" (ibid.), but I think nothing important is lost in the abbreviated version I use. Further, one might suggest that Pritchard's claim is better interpreted as one of entailment, i.e., as the claim that if one sees that *p*, then one is in a good position to know that *p*. First, let me note that some of the arguments I present against SPK's motivation would work equally well against the motivation for this alternative claim. Second, the alternative claim would leave unexplained *why* this entailment holds, whereas SPK explains this by pointing out that seeing that *p* just *is* a specific way of being in a position to know that *p*.

with SK, gives the correct verdict about the case. Pritchard claims that SPK does indeed provide the correct verdict, and explains it in the following way:

> [...] given that the defeater in play is misleading, one is *in fact* in a good position to gain knowledge of the target proposition in this case, it is just that one's inability to defeat the misleading defeater undermines one's ability to exploit this epistemic opportunity.
> (Pritchard 2012b, p. 27)

At this point, one might well feel that Pritchard has made some awkward moves. It's unclear whether Pritchard's own example can really support SPK, and it's also unclear whether the example really works against SK. I will discuss these points in turn.

4.3.1.1 Does Barnaby Support SPK?

A first thing to notice about Pritchard's example is that the phrase 'being in a good position to gain knowledge' is apparently being used in a very special sense. After all, the example was set up in such a way that Barnaby "cannot possibly know" (Pritchard 2012b, p. 27) that there is a barn before him. In that case, it certainly is not intuitive to hold that the subject is still in a good position to know the target proposition: how could one be in a good position to know that *p* if one simultaneously cannot possibly know that *p*? This tension cannot be resolved by claiming that the subject is in fact in a good position to know that *p* but unable to exploit this opportunity.

So what is Pritchard really after in his account of seeing that *p*? Pritchard explains being in a good position to know that *p* as being in a scenario that is *objectively* epistemically good (Pritchard 2012b, p. 31). Such a scenario is one in which the environment does not contain many fake look-a-likes, deceivers are not present, perceptual faculties are working properly, etc. The idea is that an agent in an objectively epistemically good scenario "is reliably forming her perceptual beliefs in such a manner that they will inevitably be true in environments which are suitably conducive to this belief-forming process, and that the agent is in just such an environment" (Pritchard 2012b, p. 29). These scenarios should be contrasted with scenarios that are *subjectively* epistemically good, where one has no defeaters for one's beliefs and correctly bases one's beliefs on the available evidence (Pritchard 2012b, pp. 30–32).

Pritchard analyzes his own example as one in which Barnaby is in an objectively epistemically good, but subjectively epistemically bad scenario. Both objective and subjective epistemic goodness matter for knowing that *p*, but only objective epistemic goodness matters for seeing that *p*. Thus, Barnaby sees that there is a barn before him without knowing that there is a barn before him. In contrast, Barn*ey* is in a subjectively epistemically good but objectively epistemically bad scenario, so he neither sees nor knows that there is a barn before him.

Although this way of cashing out SPK gets the cases right for JTBED, it's not clear that seeing that *p* really is reducible to a specific sort (viz., visual) of objective epistemic goodness. Consider the following variation on Pritchard's own Temp-case (Pritchard 2012a):

> **Angelic Help:** Angelica's perceptual mechanisms are such that, when functioning properly, they often lead to rather harmless perceptual illusions. Fortunately for her, she has a guardian angel who is determinate to protect her from forming false beliefs about the world by adapting the world to fit Angelica's perceptual experiences. For instance, when Angelica looks at the Müller-Lyer lines, her guardian angel actually lengthens the line that, at first, merely looked longer to Angelica, to make sure that the belief Angelica forms on the basis of her experience is actually true.

In this case, Angelica reliably forms her beliefs on the basis of her properly functioning perceptual mechanisms, but there is still something wrong. Her perceptual experiences successfully represent the world only because the guardian angel adapts the world to fit them, instead of the experiences being adapted to fit the world. The perceptual experiences thus have the wrong direction of fit, which makes it implausible to count Angelica as seeing that, say, the Müller-Lines from the example are unequal even though she is in an objectively epistemically good scenario. Angelica can thus be used against the idea that seeing that *p* is reducible to being in an objectively epistemically good scenario.

Now one way to respond to this problem is to incorporate a correct-direction-of-fit requirement into what constitutes objectively epistemically good scenarios. That might help to solve this problem, but there is a lingering worry that what constitutes objective epistemic goodness is not prior to what constitutes knowledge. That is, it seems as though objective epistemic goodness is best defined in terms of knowledge rather than the other way around: objectively epistemically good scenarios are those in which an agent would have gained knowledge if he did not have any defeaters and had based his beliefs correctly on his evidence. Given that Pritchard phrases being in an objectively epistemically good scenario as being in a good position to *know*, this might not be such a bad interpretation of his position. However, it would mean that JTBED can no longer be taken like a traditional, reductive JTB account of perceptual knowledge, but should rather also be understood as a version of a knowledge first account (Williamson 2000). We will see later that this puts Pritchard's epistemological disjunctivism in a dialectically weak position.

4.3.1.2 Does Barnaby Work Against SK?

We've seen that there are reasons to be skeptical of the idea that Pritchard's example can provide additional support for SPK, given that it's not so clear whether Barnaby really is in a position to know that *p*. Nevertheless, it might still work as an argument against SK. Given that the basis problem only arises if one assumes SK as an account of seeing that *p*, the example would then still provide JTBED with a way to avoid the basis problem.

However, as an argument against SK, the example is problematic for the following reason. It leans too heavily on the thought that if one were to treat oneself as having seen that *p*, one would count as actually having seen that *p*. Although I think it is likely that the subject in Pritchard's scenario will treat himself as having seen that *p*, this, by itself, provides little reason to think he actually saw that *p*. Compare Pritchard's example in which someone treats himself as having seen that *p* after he comes to know that he had misleading evidence to the following scenario in which someone treats himself as having *known* that *p* after he comes to know that he had misleading evidence.

A thief, Theo, believes that the diamonds he is after are in a certain place. Subsequently, his partner in crime tells him that the diamonds have been moved. While Theo's partner speaks sincerely, has good evidence for her claim and Theo has every reason to trust her, she is mistaken on this particular occasion: the diamonds are still in the same place. Although, let us suppose, Theo is unable to defeat the defeater he has just acquired, he continues to truly believe that the diamonds are in the relevant place. Theo goes ahead with the robbery and finds the diamonds in the very place he believed them to be. Now here is the crucial question: how would Theo treat his belief after he discovered that the defeater was misleading? Very plausibly, given that he held on to his belief all along, he will now also treat himself as having known all along. We can easily imagine him saying things like "I knew that the diamonds were in that place" when recounting the story later on. At the same time, it is of course overwhelmingly plausible that Theo is mistaken when he takes himself to have known all along. If so, the fact that Theo would treat himself as knowing once apprised of all the relevant facts provides little reason to believe that Theo did in fact know. Given that this is so, we also have reason to be suspicious of Pritchard's claim that, in his case, Barnaby's willingness to treat himself as having seen that *p* constitutes good reason to believe that he did in fact see that *p*.[7]

One might respond by claiming that Pritchard's argument is better viewed as an inference that starts from what would be a *correct description* of Barnaby, rather than from how Barnaby would treat himself. That is, we can infer that Barnaby did see that there was a barn because we would, in these circumstances, quite naturally describe him as having seen that there was a barn. However, even if it is natural to describe the subject as having seen that *p* (which already appears questionable), then it is still not yet clear that this is a *correct* description of the subject. What exactly is the motivation for supposing that we can correctly describe the subject in Pritchard's example as having seen that *p*? Note that a first motivation, that a subject would treat himself as having seen that *p*, is by now already off the table.

We are now entering into a linguistic debate about the meaning of "see that *p*" that has been more extensively explored by Craig French (2012). According to French, we can capture all that is intuitive about Pritchard's case by using either non-factive perceptual vocabulary (for instance, by saying that the subject sees the object in front of him *as* a barn), or simple seeing vocabulary (for instance, by saying that the subject sees a barn) (French 2012, p. 118). There simply is no additional motivation to suppose that it is also linguistically correct to say that the subject sees *that* there is a barn. In fact, according to French, there is linguistic evidence to the contrary:

[7]Thanks to Chris Kelp for providing me with this example.

when "*S* sees that *p*" is used in these contexts, it's being used in the *visuo-epistemic* sense. In these contexts, subjects are represented as *knowledgeable* about *p* on the basis of vision (French 2012, p. 123). If this is correct, then the case of Barnaby does not work against SK.

4.3.1.3 Barnaby's Revenge

Finally, there is another problem with the way Pritchard's example is set up. Suppose, contrary to what I have been arguing in the previous section, that the example of Barnaby does work to show that SK is incorrect on the grounds that one can infer whether a subject sees that *p* on the basis of how a subject would treat himself after coming to know all the relevant facts. Then SPK can be easily refuted on the same grounds by looking at the case of Barney.

Recall that Barney is in objectively epistemically bad, but subjectively epistemically good conditions. Barney looks at one of the few real barns in fake barn county, has no defeaters for believing that it is a real barn, and so believes that the object before him is a barn. Now it is likely that Barney will still treat himself as having seen that the object before him is a barn once he comes to know that he was actually looking at one of the few real barns in fake barn county. Given that, in Pritchard's case, the fact that the subject naturally treats himself as having seen that the target proposition is true is sufficient to conclude that the subject has actually seen that the target proposition is true, the same should go for this example: Barney saw that the object in front of him was a barn. However, the case was set up in such a way that Barney was not in a position to know that there was a barn in front of him. In consequence, if Pritchard's case constitutes a counterexample to SK, then the Barney case is a counterexample to SPK as well.

This means that a crucial building block in Pritchard's solution to the basis problem crumbles. Without SPK, Pritchard lacks an account that can accommodate, first, the thesis that seeing that *p* does not entail knowing that *p*, and, second, the thesis that Barney does not see (and thereby know) that *p*. Given that these are the desiderata for Pritchard's account of seeing that *p*, Pritchard will have failed to provide an adequate answer to the basis problem.

Of course it is open for Pritchard to reject his own example, and to accept SPK solely on the grounds that it supports an epistemological theory, i.e., JTBED, with great theoretical advantages. But now that we have a better grip on SPK, it's not so clear that these theoretical advantages are still present. For instance, consider Pritchard's claim that his theory can capture the key elements of internalism and externalism. One thing traditional internalists and externalists have in common is their attempt to provide a reductive account of knowledge. In contrast, we've seen that the most sense could be made of SPK if it was understood as being part of a knowledge first account. So traditional internalists and externalists will already feel that a key element of their position is no longer on the table. Now of course Pritchard need not incorporate *all* elements of internalism and externalism in his epistemological disjunctivism, but, as the theory was advertised, it was meant to incorporate at least all of the important elements (remember, JTBED was supposed to be the *holy grail* of epistemology). Given that both internalist and externalist have been trying to provide a reductive account of knowledge, this is not an element that can be easily put aside if a theory is supposed to be motivated by providing a synthesis of traditional internalism and externalism.

Moreover, as we will see in Sect. 4.4, Alan Millar (2010) has provided an alternative version of epistemological disjunctivism that can simply accept SK. This account incorporates many of Pritchard's desiderata without falling prey to the basis problem. So without separate motivation, SPK, and Pritchard's epistemological disjunctivism with it, are definitely in trouble.

4.3.2 Missing Key Elements

The previous sections have focused merely on the first problem for JTBED, i.e., the basis problem. The following few sections will focus on a second problem: although JTBED might succeed in incorporating some key elements of internalism and externalism, it fails to accommodate other, equally important, elements. One such element of internalism is that it is supposed to account for the New Evil Demon Intuition, while one such element of externalism is that it is supposed to account for the possibility of animal knowledge. Both elements are not captured by JTBED at all: it has to deny that demon victims are as justified as we are, and it also has to deny the possibility of animal knowledge. I will now explain these points in turn.

4.3.2.1 A Missing Internalist Element: The New Evil Demon

Recall that the New Evil Demon Intuition says that subjects radically deceived by an evil demon or malevolent scientist are as justified as we are in their perceptual beliefs (Chap. 2). Experientialism explained this intuition by pointing to the fact that demon-deceived subjects have the same experiences as us, and that these same experiences provide the same justification for the same perceptual beliefs. JTBED, however, has no satisfying way to account for the intuition.

According to JTBED, perceptual justification has to do with having factive reasons, namely, that one sees that *p*. But for Ned, our unfortunate subject in the New Evil Demon scenario, no such reasons are available. Ned's experiences are not connected to the world at all, but are induced by an evil demon. Ned does not see that *p*, and so has no factive reason for believing that *p*. This means that, contrary to the New Evil Demon Intuition, Ned lacks perceptual justification for his perceptual belief that *p*.

Of course it is compatible with JTBED that Ned *thinks* his perceptual belief that *p* is justified. The bad case in which one merely thinks one has a factive reason might be introspectively indistinguishable from a good case in which one actually has that reason. But, according to JTBED, this does not show that the justification is the same across these indistinguishable cases.

So JTBED cannot account for the New Evil Demon Intuition that Ned is as justified as we are in his perceptual beliefs. Given that this intuition seems to be an important part of internalism, JTBED cannot, after all, incorporate key elements of internalism. But if that is the case, then JTBED would lose an important part of its motivation.

Pritchard's response to this challenge is twofold. First, even though Ned lacks justification for his perceptual beliefs, he is still *blameless* in having them (Pritchard 2012b, pp. 42–3). Ned cannot distinguish his situation from one in which he is actually perceiving that *p*, so he cannot be blamed for believing what he does. The New Evil Demon Intuition can thus be accommodated if one takes it to be about blamelessness instead of justification.

However, this response lacks plausibility because it will fail to distinguish between cases that are intuitively different. Consider the case of Ben, who belongs to an isolated and benighted community the members of which share a common belief that thunderstorms indicate that their 20-eared deity is about to scratch its largest left ear.[8] When Ben witnesses a thunderstorm, he comes to believe that the deity is about to scratch an ear. Given the common belief of Ben's community concerning the link between thunderstorms and ear-scratchings, the belief Ben acquires in this case is reasonable in a way that renders him blameless in believing as he does. Yet still there seems to be an important difference between Ben on the one hand, and Ned on the other. Where Ben merely seems to be blameless in believing as he does, Ned actually appears to be praiseworthy in believing as he does, given the experiential evidence he has. At the least, Ned appears to be epistemically better off than Ben. By explaining the New Evil Demon Intuition in terms of blamelessness, Pritchard becomes unable to capture this apparent epistemic difference between Ben and Ned. So this first response against the New Evil Demon challenge fails.

Pritchard's second response claims that the New Evil Demon Intuition is actually not a key element of internalism. Pritchard attempts to support this claim by pointing out that what he calls the "New Evil Genius Thesis" (Pritchard 2012b, p. 38) does not follow from what he takes to be the important core of internalism, "Accessibilism" (Pritchard 2012b, p. 36), unless one also accepts a controversial thesis called the "Highest Common Factor Thesis" (Pritchard 2012b, p. 41). These theses are as follows:

New Evil Genius Thesis *S*'s internalist epistemic support for believing that ϕ is constituted solely by properties that *S* has in common with her recently envatted physical duplicate [or demon-deceived physical duplicate].

[8]The case of Ben is a variation of a case by Goldman (1988). See Kelp (2011), Bird (2007), and Madison (2013) for similar arguments that blamelessness and justification come apart.

Accessibilism *S*'s internalist epistemic support for believing that ϕ is constituted solely by facts that *S* can know by reflection alone.

Highest Common Factor Thesis The only facts that *S* can know by reflection alone in [an epistemically] good case are facts that *S*'s physical duplicate in a corresponding bad case can also know by reflection alone.

Although Pritchard appears to be correct in holding that the New Evil Genius Thesis does not follow from Accessibilism without accepting the controversial Highest Common Factor Thesis, it's not clear that this will help to answer the challenge posed by the New Evil Demon Intuition for two reasons.

First, as Declan Smithies (2013) notes, one can distinguish between a weak and strong variety of Accessibilism. The weak variety holds that one can know by reflection that one has rational support for believing that *p* whenever one has rational support for believing that *p*, while the strong variety holds that one can know by reflection that one lacks rational support for believing that *p* whenever one lacks rational support for believing that *p*. It's not at all clear that traditional internalism only accepts, as Pritchard suggests, the weak variety of Accessibilism. According to Smithies, "the usual internalist idea is that one is *always* in a position to know whether or not one has a certain kind of rational support for one's beliefs, and hence one is always in a position to take responsibility for one's beliefs, whether one is in the good case or the bad case" (Smithies 2013, my italics). But if that is correct, then JTBED is not able to incorporate a key element of internalism after all.

Second, as Brent Madison (2013) has pointed out, Pritchard seems to get the order of explanation the wrong way around. The New Evil Demon *Intuition*, that demon victims are as justified as we are, has independent support deriving from our response to the thought experiment. The New Evil Genius Thesis is meant as the best explanation of this intuition, from which internalists can then argue to Accessibilism. Although internalists perhaps need not argue in favor of Accessibilism via this route, it looks like a promising way to go. Moreover, the other route, from Accessibilism to the New Evil Genius Thesis, does not provide a good explanation of the New Evil Demon Intuition, as it seems too theoretical to explain why we have this response to the thought experiment (after all, even many externalists seem to share the New Evil Demon Intuition). This means that, even if we take Accessibilism rather than the New Evil Genius Thesis as the important core of internalism, then JTBED will still be left with an explanatory challenge that traditional internalists did not face in accommodating the New Evil Demon Intuition.

Given that JTBED is unable to account for the New Evil Demon Intuition, it fails to incorporate what is plausibly taken to be an important internalist element. This puts it in a dialectically weak position, as it was motivated by the thought that it could incorporate all important internalist and externalist elements. In fact, the situation is even worse, because JTBED is also unable to account for an important externalist element: animal knowledge.

4.3.2.2 A Missing Externalist Element: Animal Knowledge

Although JTBED is portrayed as being able to incorporate key elements of externalism, there is at least one key element to externalism that it cannot accommodate. This is the externalist idea that cognitively unsophisticated believers, like animals and small children, are capable of having knowledge, even when they are not cognitively sophisticated enough to have access to reasons for their beliefs. Externalist have stressed that internalist accounts of knowledge, especially those *accessibilist* accounts that require reflectively accessible reasons, face the problem of "hyper-intellectualization" (Burge 2003, p. 503, see also e.g. Dretske 1981): by demanding overly high cognitive capacities for knowledge, animal knowledge becomes impossible. The externalist solution to this problem is to give an account of knowledge that does not require justification (e.g., Goldman 1967; Dretske 1981; Kornblith 2008), or to allow for a kind of justification which does not require any access to reasons (e.g., Goldman 1979; Lyons 2009).

Since JTBED requires as an accessible reason for the perceptual belief that *p* the fact that *S* sees that *p*, it runs head-first into the hyper-intellectualization problem. What is required to access the factive reason that *S* sees that *p* is at least a belief on the part of the subject that he sees that *p*. Yet unsophisticated cognizers precisely lack the capacity for these kinds of higher-order thoughts, and so, for these believers, the relevant factive reasons for perceptual knowledge and justification turn out not to be accessible. In consequence, unsophisticated believers are unable to have perceptual knowledge and justification, and so Pritchard cannot accommodate the important externalist insight we started with.

It is not hard to see that Pritchard faces a worry about hyper-intellectualization because he embraces an accessibilist version of internalism.[9] Given that this is so, one might think that the problem can be easily avoided. After all, there are other versions of internalism on the market. Most notably, there is *mentalism*, according to which the justificatory status of one's beliefs supervenes on one's mental states (Feldman and Conee 2001). Even unsophisticated agents can host a variety of mental states, including mental states of the kind that, according to mentalism, serve to provide justification for perceptual beliefs in adult human beings. As a result, it would seem that, unlike accessibilism, mentalism can steer clear of the hyper-intellectualization worry. So couldn't Pritchard avoid the problem by abandoning accessibilism in favor of a mentalist version of internalism?

JTBED could be combined with this kind of mentalism. In fact, Pritchard himself considers this kind of view. The core idea here is that the relevant factive reasons are taken to be factive mental states in their own right. For instance, seeing that *p* would then be taken as a mental state in its own right. Moreover, it is quite plausible that a mentalist version of JTBED would solve the hyper-intellectualization problem. After all, it is plausible that even unsophisticated agents can and often do see that *p*. If so, it is also plausible that they can have perceptual knowledge that *p* in virtue of seeing that *p*.

[9]For defences of accessibilism, see (e.g. Chisholm 1977; BonJour 1985; Steup 1999).

As Pritchard himself realizes, however, he is committed to a specifically accessibilist version of internalism (Pritchard 2012b, p. 41) because of the way he motivates the internalist component of his view:

> [E]pistemic externalism entails that there is a significant degree of reflective *opacity* in the epistemic standing of our beliefs. This generates a fundamental difficulty for epistemic externalist positions, which is that it is hard on this view to capture any adequate notion of epistemic responsibility. For if the facts in virtue of which one's beliefs enjoy a good epistemic standing are not reflectively available to one, then in what sense is one even able to take epistemic responsibility for that epistemic standing?
>
> (Pritchard 2012b, p. 2)

According to Pritchard, the reason why externalism fails is that it cannot offer an adequate account of epistemic responsibility. This, in turn, is because it would seem that no account of epistemic responsibility can be satisfactory unless it features an accessibility condition. It is not hard to see that if this argument works against externalism, it will be equally effective against mentalism. Or, to be more precise, it is equally effective against any version of mentalism that would solve the hyper-intellectualization worry. After all, any such account will have to abandon the abovementioned accessibility requirement. As a result, while adopting mentalism in addition to accessibilism is an option for Pritchard, replacing the latter by the former isn't.

Another response to the hyper-intellectualization worry one might give on behalf of JTBED would appeal to Pritchard's restriction to paradigm cases of perceptual knowledge. Recall that Pritchard explicitly claims that his account of perceptual knowledge is meant for paradigm cases of perceptual knowledge (Sect. 4.2). If we restrict paradigm cases of perceptual knowledge to cases of adult human perceivers, animal knowledge will not constitute a problem for JTBED as it has been proposed by Pritchard.

However, this move would definitively be ad hoc. Pritchard characterizes paradigmatic perceptual knowledge as "perceptual knowledge which is gained in good+ cases" (Pritchard 2012b, p. 37). A "good+ case", in turn, is defined as a case with the following five characteristics: (i) The agent's environment is epistemically hospitable and the relevant faculties producing the belief that *p* are functioning properly (the case is objectively epistemically good). (ii) The agent has no defeaters for *p* (the case is subjectively epistemically good). (iii) The agent has a veridical experience and a true belief that *p*. (iv) The agent sees that *p*. (v) The agent knows that *p* (Pritchard 2012b, p. 29). Now the problem for Pritchard is that cases in which unsophisticated agents acquire perceptual knowledge often exhibit all five characteristics. If so, by Pritchard's lights, these cases qualify as paradigm cases of perceptual knowledge. Since these agents still don't satisfy the accessibility condition, the envisaged response on behalf of JTBED remains unsuccessful.[10]

[10]But couldn't Pritchard maintain that seeing that *p* features an accessibility condition in the sense that one sees that *p* only if one is in a position to know by reflection alone that one sees that *p*? If so, cases of perceptual knowledge by unsophisticated agents are not cases of seeing that *p* and so do not qualify as paradigm cases of perceptual knowledge. Again this won't do. First, there is little hope that Pritchard's motivations for an accessibility condition on knowledge and justified belief

This shows that JTBED is unable to accommodate a key element of externalism, that of animal knowledge. Given that it is also unable to accommodate the New Evil Demon Intuition, JTBED cannot be motivated by the thought that it coherently brings together all key elements of internalism and externalism. At best, it provides a combination of an internalist (accessibility) and an externalist (factivity) element, but this is not sufficient to justifiably call it the "holy grail" of epistemology.

4.3.3 An Account of Access

The last section ended by showing that JTBED cannot accommodate the possibility of animal knowledge, because animals are unable to access factive reasons in the absence of a capacity for higher-order belief. In fact, once one starts considering what it takes to access a factive reason, another problem shows up for JTBED. Pritchard does not really explain what it means to have reflective access to a factive reason, other than that it "usually means that the subject can come to know through reflection alone that she is in possession of this rational support" (Pritchard 2012b, p. 13). But if reflective accessibility is cashed out in terms of knowledge, then it seems reasonable to ask what the rational support of *this* knowledge in turn is. And now JTBED is either (a) locked in a regress, each level requiring reflective accessibility to, i.e., possible knowledge of, another reason, (b) forced to acknowledge some kind of immediate ground for the knowledge that *S* sees that *p*, or (c) forced to reject any ground for this knowledge.

I take it that option (a) is simply not palatable (*pace* e.g. Klein 1998). Accepting option (b) is also problematic. For instance, suppose that one accepts option (b) and claims that experience provides the immediate ground for the knowledge that *S* sees that *p*, in the sense that a subject having an experience that *p* would be immediately justified in believing that *S* sees that *p*. We would then have a theory according to which experience that *p* provides the rational support for the knowledge that *S* sees that *p*, which in turn provides the rational support for *p*. Surely one should then cut out the middle man, and just go for the theory which claims that experience rationally supports the belief that *p* directly. This strategy seems applicable for any ground JTBED could come up with: why not use that theory of immediate grounds directly for perceptual beliefs?

---

will work for seeing that *p* as well. Even if knowing and believing justifiably feature a responsibility condition and hence, according to Pritchard, an accessibility condition, there is little reason to think that seeing that *p* also features a responsibility condition. As a result, there is no reason to think that seeing that *p* will feature an accessibility condition. Second, as was already indicated, it is independently plausible that unsophisticated agents may and often do see that *p*, e.g. when they acquire visual perceptual knowledge that *p*. In addition, then, there is positive reason to think that seeing that *p* does *not* feature an accessibility condition.

Option (c) suffers from the same problem. Suppose one claims that one knows that one sees that *p* because of the reliability of the introspective mechanism that gives this belief as an output. Then one could ask why the reliability of introspection is sufficient for knowledge of the fact that one sees that *p*, but the reliability of perception is not sufficient for knowledge of the fact that *p*. Again the problem is that for any proposal the question will arise why this could not be used directly to account for perceptual knowledge itself.[11]

JTBED thus seems hard pressed to provide an account of reflective access to factive reasons that does not make its own theory of perceptual justification superfluous. But without such an account of access, JTBED does not even get off the ground to begin with. So this is definitively a major problem for the theory.

4.4 Millar's KFED

We have seen that Pritchard's version of epistemological disjunctivism, JTBED, encounters several severe problems that make the view unsuccessful as an account of perceptual knowledge and justification: first, the basis problem; second, the problem of missing internalist and externalist elements; and third, the problem of access. In this section, I therefore turn to Millar's knowledge first version of epistemological disjunctivism (KFED), starting with his thoughts on perceptual knowledge, and continuing with his ideas about perceptual justification.

4.4.1 Knowledge First

The crucial difference between Millar's version of epistemological disjunctivism and that of Pritchard is that Millar takes *knowledge* to be the primitive notion in terms of which justification should be explained. Millar thus follows Timothy Williamson (2000) in providing a *knowledge first* epistemology. One of the main reasons for taking this route is that it best reflects the way in which the concept of knowledge is applied in our everyday practices. We do not seem to apply the difficult conditions that have been proposed as analyses of knowledge when we assert that *S* knows *p*. What we do is different:

> We happily count people as knowing that something is an *F* when they see an *F*, and they may be presumed to have what it takes to tell of something they see that it is an *F* from the way it looks. [...] The conceptual level at which we encounter the perceptual knowledge

[11]Note that once one does use the (b) or (c) proposal to account for perceptual knowledge directly, the hyper-intellectualization problem will also be solved. This points to the fact that JTBED is *not* better off, and perhaps even worse off, than some of the more traditional internalist and externalist theories of justification.

> that we have, or that others have, is that of knowing that *p* through seeing or otherwise perceiving that *p*, by means of an ability to tell that such a thing is so from the look or other appearance of what is perceived.
>
> (Millar 2010, pp. 133–4)

According to Millar, we should not try to give a reductive account of knowledge, but should rather attempt to elucidate it by means of an investigation of the specific abilities whose exercise allow us to gain it, as it is precisely these abilities that we seem to latch on to when we ordinarily ascribe knowledge to someone.

In the case of perception, the relevant abilities are what Millar calls "perceptual-recognitional abilities", which are ways of telling that things are so from their appearances, where appearances are just the way things look, sound, smell, etc. Note that Millar does not take these appearances to serve as *evidence* on the basis of which subjects conclude that, say, something is an *F*—all the better, because the previous two chapters have focused on difficulties that are associated with such a view. Rather, the appearances are the things that our perceptual-recognitional capacities latch on to in producing the belief that something is an *F*. For this to work though, it is important that the presented appearances are *distinctive* of the recognized objects in the sense that "[w]hen an appearance of something is distinctive of *F*s, not easily could something have this appearance and not be an *F*" (Millar 2010, p. 125).

Perceptual-recognitional abilities are thus, on Millar's picture, dependent on the environment in an import way. This accords with the intuition that Barney cannot know that there is a barn in front of him when he is looking at the only real barn in fake barn county. In such a scenario, there are too many fake look-a-likes around that make the appearance of the real barn no longer distinctive of a real barn. Millar even goes as far as to claim that Barney, when in fake barn county, does not just fail to exercise the requisite perceptual-recognitional ability, but even lacks this ability altogether (Millar 2010, p. 126).

In contrast, a subject fails to exercise a perceptual-recognitional ability that he does possess in the situation where the environment is in fact favorable, but the subject nevertheless does not recognize something for what it is. This might happen when a subject is careless in his judgement, or is just unlucky enough to encounter the only fake look-a-like in the entire environment. In any case, the important point is that, as Millar construes it, the notion of exercise of an ability is a *success notion* (Millar 2010, p. 125): one cannot exercise an ability to ϕ unless one ϕs. Applied to the case of perceptual-recognitional abilities this means that one cannot exercise a perceptual-recognitional ability without *knowing* (instead of merely justifiably believing) that an object is *F*.

4.4.2 Perceptual Justification

It should be clear that Millar thinks that perceptual knowledge is not to be reduced to beliefs that are justified on the basis of some kind of evidence. But this raises a question as to what makes his view epistemological disjunctivist. Consider again what Pritchard takes to be the main thesis of epistemological disjunctivism:

> In paradigmatic cases of perceptual knowledge an agent, *S*, has perceptual knowledge that ϕ in virtue of being in possession of rational support, *R*, for her belief that ϕ which is both *factive* (i.e., *R*'s obtaining entails ϕ) and *reflectively accessible* to *S*.
>
> (Pritchard 2012b, p. 13)

Since Millar does not agree with Pritchard that an agent has perceptual knowledge *in virtue of* being in possession of factive rational support, Millar's theory would not count as epistemological disjunctivist according to this definition. But Millar does agree that an agent often has factive, reflectively accessible rational support in paradigmatic cases of perceptual knowledge:

> [E]pisodes in which I come to know that an animal I am looking at is a zebra are, barring rare, dire confusion, or rare and bizarre deception, episodes in which the fact that I see that the animal is a zebra is available to me as a reason to believe that it is a zebra and to continue to believe that it was thereafter. The intimate connection between perceptual knowledge and justified belief is accommodated by acknowledging that the fact that I see that the animal is a zebra can constitute a reason I have to take it to be one. But instead of explaining the knowledge as, so to speak, built up from justified belief, we treat the knowledge as what enables one to be justified in believing.
>
> (Millar 2010, p. 139)

On Millar's account, justification has to do with being in possession of clinching reasons for belief, i.e., reasons that settle it that the belief is true. Possession of such reasons in turn implies that a subject "[stands] in some relation to a (distinct) consideration in view of which [the subject] is justified" (Millar 2010, p. 112). Thus, even though Millar does not think that perceptual knowledge must be built up from evidentially supported beliefs, he does take perceptual *justification* to consist in having accessible, clinching reasons for belief. In paradigmatic cases of visual perceptual knowledge, these reasons are constituted by the fact that *S* sees that *p*, and this is what makes his view a version of epistemological disjunctivism. Subjects in epistemically good cases have access to factive reasons that are not available to subjects in epistemically bad cases, even though the bad cases might be introspectively indistinguishable from the good cases.

Millar even goes further than Pritchard in explaining *how* subjects in good cases have access to factive reasons. Subjects are able to *know* that they see that *p* by exercising a *higher-order* recognitional ability (Millar 2010, p. 139). These higher-order recognitional abilities differ from perceptual-recognitional abilities in that they do not latch on to the appearances of objects. After all, no object has a look that is distinctive for being seen by me. In this case it rather is my having a certain experience that is distinctive of an object's being seen by me (Millar 2011, pp. 339–40). But in other respects higher-order recognitional abilities perfectly parallel first-order recognitional abilities, and they can thus be used to explain access, and with it, justification, in terms of higher-order knowledge.

4.5 KFED's Solutions and New Problems

Although KFED is able to solve most of the problems that JTBED encountered, it also has some specific problems of its own. I will begin this section by focusing on the solution to the previous problems, and then go on to discuss the new problems that arise for KFED.

4.5.1 Solving the Problems of JTBED

Millar's KFED has some clear benefits over Pritchard's JTBED. First of all, the basis problem does not even get off the ground. KFED does not maintain that the fact that *S* sees that *p* is the evidential support in virtue of which *S* knows that *p* (and so should not presuppose that knowledge). Rather, knowledge of the fact that *S* sees that *p* puts *S* in possession of a reason that settles it whether *p* is the case. But this certainly does not preclude that seeing that *p* simply is a specific way of knowing that *p*.

Millar responds to the case of Barnaby, in which Barnaby supposedly sees that *p* without knowing that *p* because of a misleading defeater, by simply denying that Barnaby actually sees that *p* (Millar 2011, pp. 335–6). True, Barnaby might see the object in question and its relevant properties, and might even see the object *as* having those properties (Millar 2010, p. 111, n. 11), but this does not amount to the subject's seeing *that* the object has those properties. The fact that a subject might take himself to have seen that *p* once he learns that his defeater was misleading can hardly count as a strong objection to Millar's theory. So JTBED's basis problem really is no problem for KFED.

Second, KFED does not face the problems of failing to incorporate key elements of internalism and externalism. Although it is true that KFED also cannot account for the New Evil Demon Intuition because factive reasons are simply not available in the New Evil Demon scenario, KFED was not supposed to be motivated by the idea that it is a perfect synthesis of internalism and externalism. This means that, although the New Evil Demon Intuition remains problematic for KFED, it is less problematic than it is for JTBED.[12]

With regard to the externalist element that JTBED failed to incorporate, namely that of animal knowledge, KFED is even better off. KFED simply does not face a problem of hyper-intellectualization, because it does not require access to (higher-order) reasons in order to have perceptual knowledge. It only requires the exercise of perceptual-recognitional abilities, which is something that animals and small

[12]Note that I will point to a more general problem with the factivity of justification in the next section.

children are also capable of doing. This means that it is possible for cognitively unsophisticated believers to see that, and thereby know that, such-and-so is the case.

Third, KFED provides a clear account of access that is not available to JTBED. According to KFED, access to factive reasons is provided by higher-order recognitional abilities: one (usually) *knows* that one sees that *p* because one recognizes that one sees that *p*. Accessing a factive reason thus comes down to knowing that the reason obtains in a way that parallels the way in which we are able to have perceptual knowledge. This nice parallel between introspective knowledge and perceptual knowledge was exactly what JTBED was unable to accomplish.

4.5.2 New Problems for KFED

Despite the good result with regard to JTBED's problems, KFED faces a couple of problems of its own. More specifically, Millar's account has the strange consequence that justified belief turns out to be logically stronger than knowledge, while intuitively it is logically weaker. That is to say, on Millar's account, knowledge does not entail justified belief, while justified belief does entail knowledge, while intuitively it is the other way around: knowledge entails justified belief but justified belief does not entail knowledge. In what follows, both parts of this untoward consequence of Millar's account will be presented in turn.

4.5.2.1 Knowledge Does Not Entail Justified Belief

To see why, according to Millar, knowledge does not entail justified belief, just consider cognitively unsophisticated believers again. Although KFED can accommodate the possibility of animal knowledge, it cannot accommodate what one might call 'animal justification' for visual perceptual beliefs. Such justification requires that a subject be able to access the fact that she sees that *p*, which requires a higher-order recognitional ability. Now, even though unsophisticated believers might have perceptual-recognitional abilities, it's implausible that they also have higher-order recognitional abilities. This means that KFED does succumb to the hyper-intellectualization objection with regard to the *justification* of unsophisticated believers. Unsophisticated believers are *never* justified in their beliefs that *p*—even if they do know that *p*.

A similar scenario of knowledge without justification should also be possible for adult human subjects. Given that there are two distinct recognitional abilities at play in providing respectively knowledge and justification, it should be possible that the lower-order perceptual-recognitional ability is successfully exercised while the higher-order recognitional ability is not. The chicken-sexer case might be used as an instance of this possibility. Chicken-sexers, as used in epistemology, supposedly are able to reliably recognize the sex of a chick even though they have mistaken beliefs about what enables them to do so.[13] Although the chicken-sexer knows, say, that the chick is female, because he has exercised his first-order perceptual-recognitional ability, he does not know that he knows because he fails to exercise a higher-order recognitional ability. KFED would thus have the consequence that the chicken-sexer in this scenario knows that the chick is female even though he does not justifiably believe that the chick is female. And this certainly appears to be an odd result for any theory of perceptual knowledge.

4.5.2.2 Justified Belief Entails Knowledge

Let's turn to the second untoward consequence of Millar's account: that justified belief entails knowledge. The reason Millar is committed to this is that, according to him, possession of the kinds of factive reasons required for justified perceptual belief is sufficient for knowing. For instance, in the case of the visual perceptual belief that *p*, justification requires that one see that *p*. At the same time, seeing that *p* is said to be a way of knowing that *p*. In consequence, one will satisfy Millar's conditions for justified belief that *p* only if one knows that *p*. Justified perceptual beliefs that fall short of knowledge turn out to be impossible.

By way of illustration, consider the case of Barney one more time. Recall that Barney is looking at one of the few real barns in fake barn county and acquires a true visual perceptual belief that he is looking at a barn. Although Barney's belief might fall short of knowledge due to his unfortunate environment, it surely is plausible that his belief is nevertheless justified. He appears to be applying a capacity that he has successfully used many times before, and he has no reason to doubt that this case is any different. As we have already seen, Millar has no problems with accounting for the intuition that Barney doesn't know that the structure he is looking at is a barn. If Barney doesn't know that he is facing a barn, however, then, according to Millar, he also does not see that he is facing a barn. But if he doesn't see that he is facing a barn he does not have the kind of factive reason that is required for his corresponding visual perceptual belief to be justified.

In principle, this specific problem might be avoided by rejecting SK, the thesis that seeing that *p* is a way of knowing that *p*. However, this would not fit well with the knowledge first part of KFED, which was motivated by the thought that we should focus on the specific abilities by which we know that *p* (such as seeing that *p*, remembering that *p*, etc.). Moreover, most accounts of seeing that *p* will hold at least that seeing that *p* is factive. If such factive seeing is a requirement for justification, as it is according to KFED, then having a justified perceptual belief will still imply

[13]Note that it does not appear all too important whether this example is fully correct as a description of the actual world. Even if this description is not accurate of chicken-sexers, relevantly similar examples do seem possible (cf. Pritchard 2006, p. 61).

that one's belief is true. Thus, KFED will not be able to accommodate false but justified perceptual beliefs (think of the New Evil Demon scenario), which is still problematic.

It will not come as a surprise that Millar is well aware of this problem. He addresses them in the following passage:

> [T]he notion of justified belief that figures in traditional analysis and in descriptions of Gettier cases is [...] very weak. It has everything to do with a kind of reasonableness that renders one blameless in thinking that something is so, but little to do with the kind of well-groundedness that settles that something is so and on that account entitles one to take it to be so.
>
> (Millar 2010, p. 102)

Millar's idea is to distinguish between two varieties of justified belief, a strong and a weak one. The strong variety is captured by his account of justification, whereas the weak variety is unpacked in terms of blamelessness. However, we have already seen that such a move will not be satisfactory (Sect. 4.3.2.1). By explaining intuitions about justification in terms of blamelessness, an account will become unable to distinguish between cases that are intuitively different, such as those of Ben (the member of an isolated and benighted community that believes in the relation between thunderstorms and ear-scratching deities) and Ned (the demon-deceived subject). Where the former is merely blameless in believing as he does, the latter appears to be epistemically better off. The former's beliefs have no connection to truth whatsoever, while the latter's beliefs are at least formed in a way that is normally connected to the truth.

Given that KFED has no good answer to these problems, it remains unsuccessful as an account of perceptual justification. However, both JTBED and KFED do have some important insights that can contribute to a good account of perceptual justification. The next section will focus on what I take to be these important insights.

4.6 Epistemological Disjunctivist Insights

Despite the problematic consequences of JTBED and KFED, epistemological disjunctivism seems to be right about the fact that we can, and do, appeal to factive reasons in our everyday way of justifying beliefs (Sect. 4.2). Given that access to these factive reasons requires the ability to have certain higher-order beliefs (e.g., that one sees that *p*), this shows that higher-order beliefs could have an important role to play in perceptual knowledge and justification. Specifically, an appeal to higher-order beliefs might be able to accommodate the Blindsight Intuition without using the problematic notion of experiential evidence.

Recall that the Blindsight Intuition is about the case of Bill, a blindsight subject who continues to form beliefs about what is going on in his blind spot, even though he has no experiential ground on which he bases these beliefs, nor any awareness of the reliability of his own belief-forming mechanism (Chap. 2). The intuition is that Bill is not justified in his beliefs, even if these beliefs are in fact reliably produced. Experientialists explained this intuition by pointing to the fact that Bill lacks experiential evidence for his beliefs, and that such evidence is required for perceptual justification. However, in the previous two chapters we've seen that experientialism is difficult to maintain, and a different explanation of this intuition would be a welcome alternative.

Such an alternative explanation would start from the observation that Bill lacks a higher-order recognitional ability to tell that he is seeing that such-and-so is going on in his blind spot. This means, at least on Millar's KFED, that Bill does not have *access* to a factive reason, namely that he perceives that *p*, to justify his perceptual belief that *p*. If KFED is correct, then this shows that Bill lacks perceptual justification altogether. This would provide an explanation of the Blindsight Intuition that does not appeal to experiential evidence, but rather displays the importance of having higher-order access to the source of first-order beliefs.

However, we've seen that KFED has the problem of positing an account of perceptual justification that is too strong: according to KFED, perceptual justification implies, but is not implied by, perceptual knowledge. Nevertheless, something of its way of accommodating the Blindsight Intuition can be salvaged. As long as one agrees with epistemological disjunctivists that factive reasons *can* provide a subject with a special kind of justification for perceptual beliefs, then one can hold that Bill is at least epistemically worse off than us in lacking *this* kind of factive justification. In that way, one can explain the Blindsight Intuition without making perceptual justification overall too hard to come by. On this account, the Blindsight Intuition would have to do with the kinds of perceptual justification specifically available to adult *human* subjects in paradigm cases of perceptual knowledge.

Thus, there is an important insight in epistemological disjunctivism that could be incorporated by other theories of perceptual justification in order to account for the Blindsight Intuition. Although the factive reason that one sees that *p* is, as a necessary condition for perceptual justification, too strong to work, accommodating the *possibility* of this kind of justificatory support could provide theories of perceptual justification with a way of responding to the Blindsight Intuition. Moreover, it could then also easily accommodate our ordinary way of talking and thinking about perceptual justification, one of the motivations Pritchard provided for his epistemological disjunctivism. In Chap. 6, I will develop such a view in more detail, even though the full account I offer to accommodate the Blindsight Intuition is slightly different.

4.7 Conclusion

In this chapter we have looked at two different versions of epistemological disjunctivism which both hold that perceptual beliefs are evidentially justified by accessible, factive reasons of the form "I see that *p*". The first, JTBED, holds that it is in virtue of these factive reasons that we have perceptual knowledge in paradigm cases of perceptual knowledge. The second, KFED, treats knowledge as primitive, and explains perceptual justification in terms of having access to, that is, having higher-order knowledge of, factive reasons.

JTBED faces three problems that are truly devastating to the theory. First, it does not provide a strong case for the claim that seeing that *p* can be reduced to being in a *position* to know that *p* rather than to its simply being a specific *way* of knowing that *p*, and so does not adequately solve the basis problem. Second, it cannot accommodate the New Evil Demon Intuition and the possibility of animal knowledge, thereby failing to incorporate two important internalist and externalist elements. Third, it does not and perhaps cannot explain what it means to access a factive reason of the form "I see that *p*" on pain of making its own account of perceptual justification superfluous.

KFED was able to solve most of these problems by, first, acknowledging that seeing that *p* is a specific way of knowing that *p*, second, explaining perceptual knowledge in terms of undemanding perceptual-recognitional abilities, and, third, explaining access to factive reasons in terms of higher-order recognitional abilities. However, due to KFED's requirement of factive reasons for perceptual justification, perceptual justification turned out to be far stronger than one would intuitively expect: according to KFED, perceptual justification entails, but is itself not entailed by, perceptual knowledge. This means that KFED is unable to account for the possibility of animal justification and the possibility of justified but *false* beliefs.

Nevertheless, it is an important insight of epistemological disjunctivism that factive reasons can provide an important source of justification. The insight only becomes problematic once one takes this source to be necessary to have any perceptual justification at all. This suggests that a less demanding view of perceptual justification might be on the right track. The next chapter will investigate this possibility by looking at an account of justification that allows perceptual beliefs to be justified without requiring any evidence whatsoever: process reliabilism.

Chapter 5
Process Reliabilism and Its Classic Problems

5.1 Introduction

So far we have looked at versions of experientialism and versions of epistemological disjunctivism that agree in their analysis of perceptual justification as being importantly connected to *evidence*. Where experientialism takes perceptual beliefs to be evidentially justified by experience, epistemological disjunctivism takes perceptual beliefs to be evidentially justified by factive reasons of the form "I see that *p*". Both experientialism and epistemological disjunctivism faced difficult challenges. Experientialism is confronted by a version of the Sellarsian dilemma: either the posited evidential relation between experience and belief is mysterious, or the distinctive justificatory force of experience is left unexplained. Epistemological disjunctivism makes perceptual justification too hard to come by, thereby precluding cognitively unsophisticated believers as well as demon-deceived subjects from having it.

The difficulties of these evidentialist theories suggest that we should take more seriously the possibility that perceptual beliefs need only be *non-evidentially* justified. That is why this chapter focuses on a classic account of non-evidential justification proposed by Alvin Goldman (1979): process reliabilism. This account holds roughly that, for a certain class of beliefs, a belief is prima facie justified if and only if it results from a reliable cognitive process. The important point underlying this theory is that beliefs need not be based on good evidence to be justified. Instead, being the output of a reliable process is necessary and sufficient for some beliefs to be prima facie justified—even if a subject does not know anything about the reliability of this process. I discuss this classic proposal and its motivations in Sect. 5.2.

In Sect. 5.3, I discuss the two classic problems that have been raised to show that reliability is neither necessary nor sufficient for justification. These problems are respectively the New Evil Demon Problem (Lehrer and Cohen 1983) and the Clairvoyance Problem (BonJour 1985). In addition to the Clairvoyance Problem,

H. Ghijsen, *The Puzzle of Perceptual Justification*, Synthese Library 377,
DOI 10.1007/978-3-319-30500-4_5

I also introduce Lehrer's (1990) Truetemp example, and use the case of blindsight from Chap. 2 to raise similar worries for process reliabilism. Although Goldman has provided several responses to these problems, they each have their own difficulties.

Another well-known, classic problem for process reliabilism, namely the Generality Problem (Conee and Feldman 1998), will be discussed in Sect. 5.4. Briefly put, the problem is that reliabilism has to specify which precise type of process should be reliable for a belief to be justified, as any token of a belief-forming process will fall under many different types. The trouble is that it will be difficult to specify the type of process while maintaining a fully reductive account of justification in non-epistemic terms. I argue that William Alston's (1995) response to this problem, at least for the case of perception, is plausible: the relevant type of process is the psychologically real function operative in perception. Even if it is difficult to ascertain which function this is precisely, such a worry amounts to nothing more than a general worry about underdetermination.

In Sect. 5.5, I present a novel type of process reliabilism, called *inferentialist reliabilism*, that has been defended by Jack Lyons (2009). This reliabilist account explicitly distinguishes between basic, non-inferentially justifiable, beliefs, and non-basic, inferentially justifiable, beliefs. While reliability is necessary and sufficient for the former, it is not sufficient for the latter. According to Lyons, this account can be used to answer clairvoyance-style problems by pointing to the fact that these are all about *non-basic* beliefs. However, I raise some worries for this account by pointing out that one aspect of its definition of basic beliefs, which adverts to a cognitive system's etiology, is not well motivated.

In Sect. 5.6, I then discuss a variety of process reliabilism, *proper functionalism* (Plantinga 1993; Graham 2012, 2014), that can make a motivated appeal to etiology. According to proper functionalism, what matters for justification is not de facto reliability, but rather, functioning properly with the *aim* of reliably producing true beliefs. This explains the Clairvoyance and New Evil Demon Problem by pointing out that proper function is absent in the former, and present in the latter type of case. Unfortunately though, variations of these cases remain problematic even for proper functionalism.

In Sect. 5.7, I sum up the results of this chapter and conclude that the process reliabilist view needs further development before it can be used as a full-fledged theory of perceptual justification.

5.2 Classic Process Reliabilism

In this section, I'll discuss Goldman's (1979) classic version of process reliabilism. Let's start off by making clear that Goldman, in providing his process reliabilist account, intends to provide a genuine *reductive* analysis of justification in non-epistemic terms (Goldman 1979, p. 1). It is thus importantly different from Millar's (2010) non-evidential approach (that we saw in the last chapter) in which knowledge is analyzed in terms of recognitional abilities that are themselves analyzed as ways of gaining knowledge. In contrast, the process reliabilist account will not have achieved its goal if the conditions it provides for justification appeal to notions that are themselves epistemic.

Goldman is led to *reliability* as a key notion in his account of justification by first recognizing that some causal requirements are necessary for any good account of justification (Goldman 1979, p. 9), and, second, by reviewing what is distinctive of causal processes that do, and processes that don't, confer justification (Goldman 1979, pp. 9–10). The first point is made clear by looking once more at the difference between *propositional* justification and *doxastic* justification.

Take Feldman and Conee's evidentialist definition of epistemic justification as an example that only covers propositional justification:

Evidentialism Doxastic attitude *D* toward proposition *p* is epistemically justified for *S* at *t* if and only if having *D* toward *p* fits the evidence *S* has at *t*.
(Feldman and Conee 1985, p. 15)

Even if one agrees with the evidentialist line of thinking, then it should still be clear that this notion of justification is not sufficient to turn justified true belief into knowledge.[1] After all, it is possible that a subject is in possession of evidence which strongly supports *p* (which presumably means that having the attitude of belief would fit the evidence), but believes *p* on an entirely different basis (e.g., because his horoscope said that *p*). Even if there is a sense in which the belief is justified for the subject given his evidence, the fact that the subject did not actually base his belief on this evidence also provides a sense in which the belief is not justified. The former sense is often called propositional justification (which deals with the question whether a certain belief is justified for a subject, whether or not the subject accepts it), while the latter is called doxastic justification (which deals with the question whether the actual belief of a subject is justified). Given that propositional justification is clearly not sufficient for knowledge, a good epistemological theory should (also) provide an account of doxastic justification, and that is what Goldman sets out to do.

Goldman's insight is that some sort of causal requirement is necessary if one wants to adequately account for doxastic justification. Even paradigm cases of a priori infallible beliefs, such as the belief that I am here now, can fail to be justified if they are caused in the wrong way (e.g., by being based on the testimony of a pathological liar). So doxastic justification is not just related to properties of the content of a belief, but necessarily has to do with the process by which one arrived at the belief.

The second step in Goldman's reasoning compares faulty processes which do not confer justification, like wishful thinking, guessing, faulty reasoning, etc., and

[1]Feldman and Conee recognize this and therefore also provide a notion of well-foundedness that incorporates a causal aspect in the form of a basing requirement (see Chap. 2).

good processes which do confer justification, like perception, memory, and sound reasoning:

> What do these faulty processes have in common? They share the feature of *unreliability*: they tend to produce *error* a large proportion of the time. By contrast, [...] [w]hat [good] processes seem to have in common is *reliability*: the beliefs they produce are generally true. My positive proposal, then, is this. The justificational status of a belief is a function of the reliability of the process or processes that cause it, where (as a first approximation) *reliability* consists in the tendency of a process to produce beliefs that are true rather than false.
>
> (Goldman 1979, pp. 9–10)

Let me point out that the tendency of a process to produce true rather than false beliefs should not be understood in terms of the frequency of true and false beliefs that are *actually* produced by the process (although Goldman himself believes that our concept of justification is vague on this point (1979, p. 11)). If a certain process is used only once and it delivers a false belief, then we should not conclude from this that the process is unreliable, even if circumstances somehow prevent the process from ever being used again. It certainly makes sense to think of a very reliable process that, unfortunately, delivers a false belief the one time it is actually used. This shows that the tendency to produce true beliefs should rather be thought of in terms of a counterfactual: if I were to use the process in similar circumstances, then it would mostly produce true rather than false beliefs.[2]

Goldman realizes that his notion of reliability as a tendency to produce true beliefs does not fit well with the justification provided by, for example, memory and reasoning processes (Goldman 1979, p. 13). If one starts reasoning from false premises, then one might very well end up with a false conclusion even though there was nothing wrong with the reasoning process itself. Similarly, if one memorizes a false belief, then even properly functioning memory processes will not provide a true belief when one tries to recall the original belief. Goldman therefore also introduces the notion of *conditional reliability*, where a process is conditionally reliable when it has a high truth-ratio *given* that its input-beliefs are true. For *belief-dependent processes*, that is, those processes that have beliefs among their inputs, conditional reliability will be sufficient for justification as long as the input beliefs are justified themselves. In contrast, for belief-independent processes that do not have any beliefs among their inputs it is unconditional reliability that counts. Thus, according to Goldman's classic process reliabilism, the important principles for justification are the following:

Classic Process Reliabilism If *S*'s belief in *p* at *t* results ('immediately') from a belief-independent process that is (unconditionally) reliable, then *S*'s belief in *p* at *t* is justified.

[2]It is of course difficult to spell this out precisely. Nevertheless, I think it's clear that our intuitive idea of a reliable process is not connected to the actual frequency of successes of that process.

If S's belief in p at t results ('immediately') from a belief-dependent process that is (at least) conditionally reliable, and if the beliefs (if any) on which this process operates in producing S's belief in p at t are themselves justified, then S's belief in p at t is justified.[3]

(Goldman 1979, pp. 13–14)

The implicit foundationalist background of Goldman's reliabilism shows itself in the fact that the second principle already makes use of the notion of a justified belief, which would lead to circularity were it not for the first base-clause in terms of unconditional reliability. The reliabilist picture is one on which some beliefs are epistemically *basic*, that is, justified without support from other beliefs. These beliefs are the ones that result immediately from a reliable belief-independent process. The epistemically basic beliefs can then justify other beliefs through conditionally reliable processes, like inference or memory, and these beliefs in turn can then also be used to confer justification. This proposed structure of justification is a foundational one. Where the reliabilist theory differs from classical foundationalism is in its view on (a) the factor in virtue of which a belief is basic, and, relatedly, (b) the type of beliefs that can be basic beliefs.

According to reliabilism, a belief is basic when it results from an unconditionally reliable belief-independent process, whereas classical foundationalism takes beliefs to be basic when they are incorrigible or indubitable.[4] Given this commitment to incorrigibility or indubitability, the type of basic beliefs that classical foundationalism appeals to are usually mathematical beliefs or beliefs about one's own experiences. In contrast, the type of beliefs that reliabilism takes to be basic are just those that follow from reliable belief-independent processes. These are not just, say, mathematical and introspective beliefs, but also *perceptual* beliefs. The benefits of such an account over classical accounts of foundationalism should be clear: one does not have to cross the seemingly unbridgeable gap between beliefs about one's own experience and beliefs about the world. Instead, beliefs about the world can themselves be as epistemically basic as beliefs about one's own experiences. What's more, the criterion of incorrigibility or indubitability is insufficient to provide a plausible theory of justification for the reasons mentioned above: although this might be used to define propositional justification, more is needed for doxastic justification. One could hold incorrigible beliefs for bad reasons, making them unjustified despite their incorrigibility. So a reliabilist foundationalism is certainly an improvement over the classic version of foundationalism.

But let's get back to the principles Goldman introduced for reliabilist justification. Even with the distinction between belief-dependent and belief-independent processes, the stated conditions will not be sufficient for ultima facie justification. The reason for this is that counterexamples can be constructed in which a subject holds a belief that is reliably caused, but in which that same subject also has

[3]Note that the 'immediately' is probably meant to exclude tracing back beliefs to very distal causes.

[4]Take sense-datum theorists like Price (1932) as an example, see Chap. 1.

defeating evidence against the truth of this belief. Intuitively, the subject in such a scenario would not be justified in his belief because of his defeating evidence, even if the belief was the output of a reliable cognitive process. Goldman gives the following example:

> Suppose that Jones is told on fully reliable authority that a certain class of his memory beliefs are almost all mistaken. His parents fabricate a wholly false story that Jones suffered from amnesia when he was seven but later developed *pseudo*-memories of that period. Though Jones listens to what his parents say and has excellent reason to trust them, he persists in believing the ostensible memories from his seven-year-old past. Are these memory beliefs justified? Intuitively, they are not [. . .]
>
> (Goldman 1979, p. 18)

The problem for reliabilism is that it has, so far, not provided any resources to preclude Jones' beliefs from being justified. Moreover, it cannot just add that subjects should lack any defeating evidence to be justified, as evidence itself is an epistemological notion that does not fit the reliabilist's aim of providing a reductive analysis of justification in non-epistemic terms. So reliabilism has to find another way of dealing with examples like that of Jones.

Goldman's own solution makes use of the idea that "[t]he justificational status of a belief is not only a function of the cognitive processes *actually* employed in producing it [but] also a function of processes that could and should be employed" (Goldman 1979, p. 20). If Jones had properly taken into account his parents' testimony, then he would not have believed the relevant memory-based beliefs, and that is why he is unjustified. These considerations lead Goldman to the following (simplified) presentation of his theory:

> If *S*'s belief in *p* at *t* results from a reliable cognitive process, and there is no reliable or conditionally reliable process available to *S* which, had it been used by *S* in addition to the process actually used, would have resulted in *S*'s not believing *p* at *t*, then *S*'s belief in *p* at *t* is justified.
>
> (Goldman 1979, p. 20)

Defeat is thus spelled out in terms of a counterfactual: if there is a reliable or conditionally reliable process available to *S* which, had it been used, would have resulted in *S*'s not believing that *p*, then the prima facie justification for the belief that *p* is defeated.[5]

We now have reliabilist conditions for justification arising out of belief-independent and belief-dependent processes, and a reliabilist account of defeat. In the next section we look at some well-known counterexamples that attempt to show that this full reliabilist theory is still not adequate to accommodate all our intuitions about justification.

[5] I will have more to say about this defeat-condition in the next chapter.

5.3 Counterexamples to Reliabilism

5.3.1 New Evil Demon Problem

The first counterexample to reliabilism attempts to show that the reliability of a belief-forming process is not *necessary* for justification. That is, the example is one in which a subject intuitively has justified beliefs even though he lacks a reliable belief-forming process, viz. the New Evil Demon scenario. Recall Lehrer and Cohen's description of this scenario:

> Imagine that, unknown to us, our cognitive processes, those involved in perception, memory and inference, are rendered unreliable by the actions of a powerful demon or malevolent scientist. It would follow on reliabilist views that under such conditions the beliefs generated by those processes would not be justified. This result is unacceptable. The truth of the demon hypothesis also entails that our experiences and our reasonings are just what they would be if our cognitive processes were reliable, and, therefore, that we would be just as well justified in believing what we do if the demon hypothesis were true as if it were false.
>
> (Lehrer and Cohen 1983, p. 192)

As Jack Lyons (2013) points out, a first thing to notice about this example, as it is stated here, is that it actually fails to address Goldman's classic process reliabilism. Contra Lehrer and Cohen's claim, Goldman does not require memory and inference to be reliable to produce justified beliefs. Instead, these belief-dependent processes merely require, first, *conditional* reliability and, second, justification of the input beliefs. So inference and memory can still generate justified beliefs as long as the input beliefs are justified and the processes themselves remain conditionally reliable, that is, reliable in producing true beliefs given that the input beliefs are true.

Since the idea of the example is that "our reasonings are just what they would be if our cognitive processes were reliable", the example appears to assume that inference and memory are still conditionally reliable. The example is not meant as one in which it seems to subjects as if they are making valid inferences but are instead continually making invalid moves (in that case it would also be doubtful whether we would still have the intuition that the subjects are justified in their beliefs). Rather, the envisaged subjects are supposed to be psychologically identical to us, even though they are continually deceived by an evil demon. But given this psychological identity, belief-dependent processes would remain conditionally reliable and would, therefore, remain capable of producing justified beliefs as long as the input beliefs are justified. Examples of such justified input beliefs would be those delivered by introspection, or, if there is such a thing, rational intuition. These processes could remain totally reliable even if perception was not. So subjects of the above New Evil Demon scenario, or demon-worlders for short, could still have some fully justified beliefs.

Another point of Lyons (2013) is that reliabilists can also use the conditional reliability of belief-dependent processes to account for the positive epistemic status of beliefs based on memory or inference. These beliefs, even if they are not always categorically justified, could still be said to be *conditionally justified*. To make plausible that there is something epistemically interesting about this notion, Lyons gives the following example:

> Forgetting reliabilism for a moment, suppose an agent draws an impeccable inference from an unjustified belief. The resulting belief is unjustified, but to say only this is to ignore the fact that the agent has achieved some positive epistemic accomplishment by drawing a proper inference, despite the flawed premise. An agent who draws valid inferences from unjustified premises is doing something epistemically right, something that an agent who draws invalid inferences is not. Being conditionally justified in this way is a positive epistemic achievement, and having conditional justification is a positive epistemic status, even if subordinate to being justified.
>
> (Lyons 2013, p. 8)

Allowing for the notion of conditional justification gives the reliabilist a way to account for the positive epistemic status of beliefs that result from belief-dependent processes that goes beyond their mere *blamelessness*.[6] Brainwashed individuals or subjects growing up in some kind of benighted community might be blameless in their beliefs even if there is nothing epistemically positive about them. Conditional justification, in contrast, is better than that: beliefs that are conditionally justified are formed on the basis of other beliefs in an epistemically proper way.

Even if it is true that in the New Evil Demon scenario, first, the conditional reliability of memory and inference is preserved, second, the categorical justification of introspective and rationally intuited beliefs is not impugned, and third, conditional justification remains present for memorial and inferential beliefs, then we have still not dealt with what looks like the heart of the problem. What makes the New Evil Demon scenario so problematic for reliabilism is that we have the intuition that the *perceptual beliefs* of demon-worlders are fully justified *despite* their resulting from an unreliable cognitive process. The previous points can at best be taken as damage control on behalf of the reliabilist without a satisfying response to this problem.

There are two ways for reliabilists to respond to the worry about the perceptual beliefs of demon-worlders, the gist of which is already anticipated by Goldman (1979, pp. 17–8). Either one claims that demon-worlders are justified in their beliefs and defends a version of reliabilism that can accommodate that claim, or one claims that demon-worlders merely appear to be justified and presents a theory as to why we mistakenly have this intuition. The first option can be taken by making the notion of reliability relative to, e.g., the actual world (Goldman 1979), 'normal worlds' (Goldman 1986), or a subject's 'home world' (Majors and Sawyer 2005). One would then (roughly) claim that beliefs are justified only if they are the output of a cognitive process that is reliable in the actual world (or etc.), instead of being reliable in the world at which the belief is located. But such a strategy appears flawed from the beginning: why would reliability in a specific world matter for the justification of a belief in a different world?[7] For instance, suppose that there is a world where

[6] Cf. Goldman's (1988) approach, and also see the discussion of blamelessness in Sect. 4.3.2.1.

[7] Note that Majors and Sawyer do provide a rationale for their home-world-reliabilism: because the content of perceptual states and beliefs is individuated with respect to the home world in which their subjects developed, the process of going from a certain perceptual state to a certain belief

clairvoyance is a reliable process. Whether or not the beliefs of clairvoyants in that world are justified (this will be the question for the next section), it seems that the question of justification in that world just has nothing to do with the fact that clairvoyance is not a reliable process in *our* world.

The second way, which tries to explain why we would *mistakenly* have the New Evil Demon Intuition, appears to be the better route for reliabilists. As before, an appeal can be made to considerations about possible worlds, but this would now purely be used to *explain* our intuitions instead of *corroborating* them. Since perception is a reliable process in the actual world, it is not strange that we would have the intuition that demon-worlders, who also rely on perception, are also justified. We mistakenly carry over the actual reliability of perception to the demon-world when we answer the question of whether demon-worlders are justified in their perceptual beliefs.

Although this response is more plausible than making the relevant notion of reliability relative to the actual world, it still faces difficulties. The set-up of the New Evil Demon scenario is pretty clear, and it is assumed from the outset that the relevant perceptual processes are *not* reliable. So why would we then still take into account the fact that perception is reliable in the actual world when we think about the question of whether demon-worlders have justified beliefs? At least more needs to be said to make this suggestion plausible.[8]

5.3.2 Clairvoyance, Truetemp, Blindsight

Where the New Evil Demon scenario attempts to show that reliability is not necessary for justification, the examples in this section try to show that it is not *sufficient*. That is, the examples are meant to make plausible that a reliably produced belief can still be unjustified. The relevant type of example is again one we already encountered in an earlier chapter:

> **Clairvoyance:** Norman, under certain conditions which usually obtain, is a completely reliable clairvoyant with respect to certain kinds of subject matter. He possesses no evidence or reasons of any kind for or against the general possibility of such a cognitive power or for or against the thesis that he possesses it. One day Norman comes to believe that the President is in New York City, though he has no evidence either for or against this belief. In fact the belief is true and results from his clairvoyant power under circumstances in which it is completely reliable.
>
> (BonJour 1985, p. 41)

---

should be reliable in that same home world. Although this is a good start, such a strategy will not help with the Clairvoyance Problem of the next section, and it is also unable to respond to a Reversed Evil Demon scenario (for more on this, see Sect. 5.6.2).

[8]Note that I will pursue a different version of explaining (away) the New Evil Demon Intuition in the next chapter.

The question for reliabilism is this: is Norman's belief that the President is in New York City justified? According to Goldman's reliabilism, it seems that it is. After all, the belief is the result of a reliable belief-independent process, and there appears to be no alternative process available that would have led to Norman's not believing that the President is in New York City. This means that, according to reliabilism, the belief is both prima facie and ultima facie justified. Yet intuitively Norman does not appear to be justified. BonJour explains this intuition as follows:

> Norman's acceptance of the belief about the President's whereabouts is epistemically irrational and irresponsible, and thereby unjustified, whether or not he believes himself to have clairvoyant power, so long as he has no justification for such a belief. Part of one's epistemic duty is to reflect critically upon one's beliefs, and such critical reflection precludes believing things to which one has, to one's knowledge, no reliable means of epistemic access.
>
> (BonJour 1985, p. 42)

A similar example as the clairvoyance case has been proposed by Keith Lehrer (1990, pp. 163–4), although Lehrer frames his discussion in terms of knowledge. However, the example works equally well for justification:

> **Truetemp:** Suppose a person, whom we shall name Mr. Truetemp, undergoes brain surgery by an experimental surgeon who invents a small device which is both a very accurate thermometer and a computational device capable of generating thoughts. The device, call it a tempucomp, is implanted in Truetemp's head [...] [and] sends a message to his brain causing him to think of the temperature recorded by the external sensor [...]. He accepts [the thoughts about temperature] unreflectively, another effect of the tempucomp. Thus, he thinks and accepts that the temperature is 104 degrees. It is.
>
> (Lehrer 1990, pp. 163–4)

It seems that Mr. Truetemp does not know, and does not have a justified belief that, it actually is 104 °F even if this belief was caused by a reliable process. Yet Goldman's process reliabilism again seems to give the verdict that the belief is both prima facie and ultima facie justified. Lehrer's diagnosis of the situation is similar to that of BonJour: "more than the possession of correct information is required for knowledge [or justification]. One must have some way of knowing that the information is correct" (Lehrer 1990, p. 164). Again, some sort of higher-order reflection appears to be posited as a necessary condition for knowledge or justification.

The first thing to notice about the force of both these examples is that, even if we have the intuition that Norman and Truetemp are unjustified, BonJour's and Lehrer's specific diagnosis might still be mistaken. Their explanations demand higher-order capabilities that cognitively unsophisticated agents lack. Yet we also have the intuition that such agents *are* capable of having justified beliefs or at least knowledge. So Bonjour and Lehrer's own analyses are in conflict with some of our other epistemic intuitions. This means that the floor is open to reliabilists to provide a better analysis of these cases.

One possible analysis would take a similar line as before by connecting reliability, or our judgments about reliability, to the actual world. However, first, we've already seen reasons to be skeptical of such a response, and second, the Truetemp case is so close to the actual world that this reply really lacks plausibility here. The reliabilist is thus in need of a different sort of response to these cases.

Compare the case of Norman and Truetemp to the imagined case of Bill and his blindsight (Chap. 2). Bill lacks experience in a certain part of his visual field due to damage to his visual cortex, yet still reliably forms beliefs about what goes on in that part of his visual field. Bill himself has no idea where his beliefs come from, and does not know whether they are correct or not. Now, if Bill were still capable of having the relevant experiences, then surely he would be justified in his perceptual beliefs. However, lacking such experiences, he appears to be unjustified in the beliefs he forms even though they are produced by a reliable mechanism. The intuition that Bill is unjustified might thus be explained by the fact that Bill lacks experiential evidence for his belief. However, the same explanation is not as clearly applicable in the case of Norman and Truetemp. Even if Norman and Truetemp had a specific phenomenological experience associated with their respective clairvoyant and tempucomp beliefs, then they might still lack justification. So perhaps there is something to the idea that it is the lack of knowing where the beliefs come from which prevents Bill, Norman, and Truetemp from being justified after all.

Goldman tries to utilize this idea by making the lack of knowledge of the source of the beliefs responsible for a *defeater* of the otherwise justified beliefs:

> A more attractive way to bolster reliabilism is to add a weaker supplementary condition, a negative higher-order condition [...] This says that a cognizer, to be justified, must not have reason to believe that her first-order belief isn't reliably caused. This promises to handle the clairvoyance and Truetemp cases very smoothly. Surely Truetemp, like the rest of us, has reason to think that beliefs that come out of the blue—as far as one can tell introspectively—are unreliably caused. Hence he has reason to believe that his spontaneous beliefs about the precise ambient temperature are unreliably caused. So his first-order beliefs about the ambient temperature violate the supplementary condition, and therefore are unjustified. For this maneuver to help reliabilism, of course, "defeat" must be cashed out in reliabilism-friendly terms.[9]
>
> (Goldman 2011a)

The difficulty now of course consists in cashing out the notion of defeat in reliabilism-friendly terms. One might go back to the earlier mentioned anti-defeater condition, according to which a subject should not have available another reliable process which would lead him to not believe that p, and apply this to the case of Norman, Truetemp, and Bill. They do have a reliable process available that could lead them to doubt their respectively clairvoyant, tempucomp and blindsight belief, namely, reflection on the fact that their beliefs came out of the blue. Were they to do this, then they could come to the conclusion that they should not believe as they do. However, although there might be something to this suggestion, it needs serious development before it can be made to work. The notion of 'reflection' is just too vague to denote a specific alternative process that is responsible for defeat.

[9]In the passage left out Goldman refers to Goldman (1986, pp. 111–2) as the place where he introduces such a condition. But a general anti-defeater condition is already present, as we saw, in Goldman (1979).

For instance, by reflecting on a certain set of my beliefs I might be able to derive a contradiction, thereby forcing me to give up a specific belief in the set. But if the derivation is difficult, or just contains many steps, then it certainly is not obvious that my justification is defeated before the derivation is actually carried out. One could claim that this is not the type of reflection at issue, but then it should be made clear which type is.

This means that the cases of Norman, Truetemp, and Bill remain problematic for the reliabilist. Why is it that these subjects lack justification for their beliefs, even though they are the result of reliable cognitive processes? The most plausible solution seems to be one that accommodates their lack of justification in terms of defeat, but more work is needed to explain, in reliabilism-friendly terms, how their beliefs are defeated exactly.[10]

5.4 The Generality Problem

The former section presented two types of examples against reliabilism that were meant to show that reliability is neither necessary nor sufficient for justification. In this section we look at a classic problem for reliabilism which purports to show that it cannot even get off the ground as a reductive analysis of justification. This problem for reliabilism is already mentioned by Goldman when he introduces the theory:

> A critical problem concerning our analysis is the degree of generality of the process-types in question. Input-output relations can be specified very broadly or very narrowly, and the degree of generality will partly determine the degree of reliability. A process-type might be selected so narrowly that only one instance of it ever occurs, and hence the type is either completely reliable or completely unreliable. (This assumes that reliability is a function of *actual* frequency only.) If such narrow process-types were selected, beliefs that are intuitively unjustified might be said to result from perfectly reliable processes; and beliefs that are intuitively justified might be said to result from perfectly unreliable processes
>
> (Goldman 1979, p. 12)

Although it is Goldman himself who first mentions this Generality Problem for reliabilism, its full force has been developed by Feldman (1985), Conee and Feldman (1998) and Feldman and Conee (2002). The problem comes down to the fact that each token belief-forming process falls under many different types, and the reliabilist has not specified the reliability of which of these many types is relevant in determining whether a subject is justified in his belief. What makes the problem so difficult is that the reliabilist should specify the relevant type of belief-forming process while steering between the Single Case Problem, where the type is so narrow that it has only one instance, and the No Distinction Problem, where the type is so broad that beliefs of different epistemic status are produced by tokens that fall under it (Feldman 1985, p. 161).

---

[10] I will pursue this solution further in the next chapter.

An example helps to make the problem more clear. At the moment, I'm looking at the laptop in front of me and thereby know that there is a laptop in front of me. The token belief-forming process at work here can be described as an instance of the type *forming a belief on the basis of vision*. To see whether, according to reliabilism, I am justified in my belief, we would have to determine whether this type of process is reliable. Since there are many cases in which I would believe something on the basis of vision in which my belief would nevertheless be false (for instance, in dimly lit environments), it seems that the description of the process-type is too broad, i.e., it runs into the No Distinction Problem. So the type should be specified more narrowly as, for instance, *forming a belief on the basis of vision in good lighting conditions*. But this type will still be too broad, as one should also be paying attention, have an unblocked view of the object under consideration, etc.

Eventually, one might end up with the type *forming the belief that there is a laptop in front of me because of having a retinal image with such-and-such properties, while paying sufficient attention, on Wednesday, March 18, 2015, 3 PM in the actual world*. But this type is so specific that it has only one instance, and therefore will be trivially reliable because the belief happens to be true. Here the reliabilist would indeed come up with the right answer to the question of justification, but not, it seems, for the right reasons. For if all processes were individuated in this narrow way, then all true beliefs would come out as justified, and all false beliefs as unjustified. So reliabilism has to find a way to specify which type of belief-forming process is relevant in determining whether a belief is justified that steers between the No Distinction Problem and the Single Case Problem to the correct level of generality.

Notice that the problem is not just the problem of finding a type of cognitive process the reliability of which can be intuitively relevant for determining whether or not a certain belief is justified. One can probably always find a process the (un)reliability of which can be used to explain our intuitive judgments about the justification of the belief at issue. The problem is that the reliabilist has to specify a *principled* way of determining for any case which type of process should be looked at to determine whether or not the belief is justified in that case.

Now one thing that is fairly obvious is that the type *forming a belief on the basis of vision on a Wednesday* is not the type of belief-process that a reliabilist like Goldman had in mind when he proposed his analysis of justification. The natural explanation for this is that its being a Wednesday has no causal impact on the production of vision-based belief. This train of thought leads to Goldman's later suggestion that "the critical type is the narrowest type that is causally operative in producing the belief token in question" (1986, p. 50). However, Conee and Feldman rightly remark that the narrowness of this type creates problems for reliabilism:

> [This suggestion] classifies into the same relevant type only beliefs that share all internal causal predecessors. Thus, on the reasonable assumption that the content of any normally formed belief is causally determined by its antecedent psychological causes, according to [this suggestion] each relevant type can have only one content for its output belief. This makes trouble in cases in which the proposition believed dictates the truth-ratio of all process types leading only to it.
>
> (Conee and Feldman 1998, p. 14)

Examples of the propositions Conee and Feldman are thinking about are necessary truths and necessary falsehoods. If types are individuated on the basis of all causally operative items, then processes that belong to the same type also provide the same output beliefs. Hence, processes that produce beliefs in necessary truths will *always* be fully reliable, while processes that produce beliefs in necessary falsehoods will always be fully unreliable—regardless of the details of the process. And this surely is the wrong result, processes have now become too narrowly defined again.

At this point, a suggestion by William Alston (1995) might be helpful. According to Alston, the relevant type of belief-forming process is the psychologically real function that is actually operative in producing beliefs as outputs on the basis of certain inputs. In the case of visual perception, such a function would take certain features of the environment as the input and would produce certain beliefs about the environment as the output. Alston clarifies his proposal with the help of the following example:

> Consider the formation of a visual perceptual belief that a maple tree is in front of one. The input will be a visual "presentation" of a certain sort, one that involves the perceived object's *looking* a certain way. The mechanism that is activated will take account of certain phenomenal features of the presentation, while others will play no role. Certain shape features, certain color features, the spatial distribution of variously colored regions, and contrasts with the surrounding field will be "picked up" by the mechanism, while others will be ignored. As for the latter, many details of the presentation could have been different without changing the content of the belief generated.
>
> (Alston 1995, p. 13)

The idea behind Alston's proposal is that we could (and that psychology does) investigate which features of a maple tree have a bearing on the produced output belief that the tree is a maple tree, thereby coming to know more about the *unique* psychologically real function at work. However, Conee and Feldman disagree. According to them there are still numerous psychologically real function instantiated in every instance of belief formation:

> There is a very narrow function that goes from just the leaf shape that Smith notices as input to just the output of Smith's particular belief that a maple tree is nearby. There is another function, one that maps a variety of fairly similar inputs, including the particular shape that Smith noticed, onto some belief or other to the effect that there is a maple tree nearby, including the belief Smith forms. There is a broader function, one that maps a variety of somewhat similar inputs, all involving visual shapes, onto either the belief that there is a maple tree nearby or the belief that there is an oak tree nearby or the belief that there is elm tree nearby, etc. There are still broader types that include the original pair, and add new inputs involving various other sensory cues. In many cases, all these functional causal relations, and many others as well, would be actually operative in forming Smith's belief.
>
> (Conee and Feldman 1998, p. 12)

However, at this point Conee and Feldman seem to overstate their case. For instance, it seems implausible that the narrow function going *just* from leave shape to the belief that there is a maple tree nearby is actually operative in many cases of maple tree belief formation. If it was, then Smith would form his beliefs about maple trees independently of all kinds of other factors such as the color and texture of the leaves under consideration. And surely no one's *perceptual system* uses those kinds of simplistic identification methods. Conee and Feldman do raise a challenging question though: should we think of the function going from leaf shape to maple tree identification as one that goes from a specific number of points of the leaf (7) to the output belief that it is a maple tree, or as one that goes from a range of number of points to a range of belief outputs (e.g., that it is a maple tree, that it is an oak tree, etc.)? Their own answer, that both functions would in many cases be actually operative in belief formation appears implausible as one might surely be able to recognize maple trees without recognizing oak trees. And if one was able to recognize both types of trees, why then suppose that both functions would be operative in a specific case of recognition instead of just one?

One thing that is, of course, true, is that it is difficult to establish which precise function is utilized by the perceptual system. For instance, should we think of attention as somehow implementing a different function from input to output, or as a process that influences the input to the same perceptual-identification function? But the fact that these questions might be difficult to answer does not show that there is *no* answer to them. So even though Conee and Feldman raise a challenge to determine which precise function is at work in a case of belief formation, they do not make plausible that there actually are several functions at work that each have a different reliability. And this is what they have to show if they want to make the Generality Problem insurmountable for reliabilists with regard to perceptual justification.

Conee and Feldman still succeed in creating difficulties for reliabilism for a different reason though. Even if it is true that psychology could eventually come up with the actually operative functions at the right level of generality, the reliability of which determines whether a belief is justified or not, then it is still highly unlikely that we are currently estimating the reliability of precisely those functions when we attribute knowledge or justified beliefs to someone. Of course this does not imply that justification is not reducible to reliability (water is H_2O even though we do not normally judge a liquid to be water by checking whether its molecular structure is constituted by hydrogen and oxygen atoms), but it could lend at least some extra credence to plausible alternatives to reliabilism. However, I will not pursue this type of objection to reliabilism any further here.

5.5 Inferentialist Reliabilism

So far we have looked at Goldman's classic process reliabilism and the problems it encounters, i.e., the New Evil Demon Problem, the Clairvoyance Problem, and the Generality Problem. In this section we look at a version of process reliabilism, developed by Jack Lyons (2009), that aims to solve at least some of these problems. One of the most important features of this *inferentialist reliabilism* is its focus on distinguishing between basic beliefs, that require only reliability for their justification, and non-basic beliefs, the justification of which requires something different than mere reliability (viz. conditional reliability and justification of the beliefs they are based on). This distinction is based on the intuitive thought that we, as human agents, simply cannot have immediate justification for certain complex beliefs without additional evidential support from other beliefs. Take, say, the belief that externalism about perceptual justification is correct, or the belief that "bats are more closely related to primates than to rodents" (Lyons 2009, p. 135). Intuitively, we need more than a mere reliable process to make such beliefs justified, we need background beliefs from which we can then justifiably infer those beliefs. On the other hand, there are also beliefs that intuitively are capable of being immediately justified without any additional evidential support from other beliefs, such as the belief that there is now a table before me. The latter belief would be epistemically basic, whereas the former would be epistemically non-basic.

Lyons explicitly presents his account as foundationalist: basic beliefs provide the foundation from which non-basic beliefs can be justifiably inferred if certain conditions hold. This explicit focus on the distinction between basic, non-inferential, beliefs and non-basic, inferential beliefs is what leads Lyons to call his view 'inferentialist reliabilism'. Now, it is true that Goldman also makes a distinction that could be used to define basic and non-basic beliefs. Recall that Goldman has separate conditions for the justification of beliefs that result from belief-independent and belief-dependent processes, where beliefs formed by belief-dependent processes require, not reliability, but conditional reliability and justified input beliefs. The basic beliefs would then be the beliefs formed by belief-independent processes, while the non-basic beliefs would be the beliefs formed by belief-dependent processes.

However, as Lyons points out, this suggestion leads to several problems (Lyons 2009, pp. 125–7). The epistemic basicality of beliefs has to do with the question of whether these beliefs *evidentially* depend on other beliefs, i.e., whether these beliefs are based on other beliefs. But this is not captured by defining basic and non-basic beliefs in terms of being the result of respectively belief-independent and belief-dependent processes. For instance, memory and introspection both count as belief-dependent processes, but memorial and introspective beliefs do not seem to be *based* on the beliefs that serve as the input for memory and introspection. This is clearest in the case of introspection: if I use introspection to find out whether I believe that *p*, then I will still be justified in believing that I believe that *p* even if my belief that *p* itself is completely unjustified. Given that one can have a justified higher-order belief even if the first-order belief is itself unjustified, it seems that this first-order belief does not *evidentially* justify the higher-order belief. After all, it is plausible that successful evidential justification requires that the evidential justifier is itself justified.[11]

Lyons therefore presents an alternative account to answer the question of which beliefs are basic (what he calls the 'Delineation Problem' (Lyons 2009, p. 18)):

> [...] [B]asic beliefs are those that are the outputs of a certain kind of operation of a certain kind of cognitive system, in particular, an inferentially opaque cognitive system, which results from learning and/or innate developmental processes, where the system is not basing its outputs on any doxastic inputs.
>
> (Lyons 2009, p. 166)

[11] See Sect. 2.3 for more on the distinction between evidential and non-evidential justification.

To make this account more clear, we should first look at Lyons' notion of a cognitive system. According to Lyons, a cognitive system for task *T* "is a virtual machine that does *T* and nothing else, is self-sufficient with respect to *T*, and is such that the *T* it does is a functionally cohesive task, rather than some gerrymandered collection of independent tasks" (Lyons 2009, p. 90). The idea is that while there is no such thing as, say, a coffee-cup-identification system (the virtual machine that identifies coffee-cups also identifies other objects), or an auditory-and-gustatory system (not functionally cohesive), there *is* such a thing as an object-identification system. Note that a cognitive system might consist out of smaller subsystems, each carrying out its own subtask (Lyons 2009, pp. 90–91).

Furthermore, some of these cognitive systems are *inferentially opaque*. An inferentially opaque system provides doxastic outputs that are "cognitively spontaneous in BonJour's (1985) sense: they are not the result of an introspectible train of reasoning from earlier beliefs" (Lyons 2009, p. 95). This inferential opacity should be distinguished from introspective opacity, which just means that there is no conscious accessibility to intermediate representations in a system's computing output from input (Lyons 2009, p. 95). A cognitive system might be inferentially opaque without being introspectively opaque. The visual system is a good example: although we seem to be aware of some of the *non-doxastic* representations that are used by the visual system in generating its output visual beliefs, we are not aware of any train of reasoning leading to these visual beliefs. Even if the visual system actually reasons (in some sense of the term) from a certain retinal input to conclusions about the environment, we are certainly not aware of this reasoning.

Finally, Lyons also takes some cognitive systems to be able to *base* their output on doxastic inputs (Lyons 2009, pp. 138–40). According to Lyons there is a difference between the way introspection uses beliefs as input to arrive at higher-order beliefs about them, and the way that, say, an "AND-elimination system" (if there is such a thing) might use the input belief "p & q" to arrive at the conclusion "q". In the former case the system does not use the input belief *as evidence* for its output belief, since "to take a belief as evidence for another is to take the content of the first as evidence for the content of the second, and this certainly isn't the case with belief introspection" (Lyons 2009, p. 140). Although Lyons seems to be right in his claim about what it means to take a belief as evidence for another, it also seems difficult to establish whether a cognitive system is actually doing that. However, since we are primarily concerned with perceptual beliefs and perceptual systems that do not seem to use doxastic inputs as evidence at all, this will not be a problem for our discussion.[12]

Now that we have a better grasp of Lyons' general account of basic beliefs as those beliefs that are the outputs of non-inferential operations (i.e., operations that do not base the outputs on doxastic inputs) of inferentially opaque cognitive systems

[12]Cognitive penetration of beliefs on perceptual experience would, on this account, have to be interpreted as a causal rather than evidential process (see e.g. Lyons (2011) and Siegel (2012) for discussion of cases of cognitive penetration). But it's not really clear that this is problematic.

that developed in the right way, we can ask which specific beliefs count as basic. This question will depend on our precise cognitive architecture, and so can only be fully answered by cognitive science. For instance, if the beliefs we have about chairs, tables, laptops, etc. turn out to depend on the inferential operation of our perceptual system, i.e., if the perceptual system bases these beliefs on lower-level beliefs about shapes, colors, etc., then these beliefs will not be basic. However, Lyons thinks it's likely, given what we already know about our own cognitive architecture, that the beliefs we intuitively take to be perceptual are also basic: after a certain amount of learning beliefs about objects around us will be the result of the non-inferential operation of the perceptual system (Lyons 2009, p. 111). There can even be differences in basic beliefs between two observers: for the expert, the belief that there is a copperhead snake might be basic, while it could be non-basic for the novice. This all has to do with the fact that our perceptual system develops: by learning we can come to identify objects non-inferentially even though we started out identifying them inferentially.

By incorporating the distinction between basic and non-basic beliefs in the classic reliabilist theory, Lyons proposes the following novel version of process reliabilism (limited to what is important for perceptual justification):

Inferentialist Reliabilism If *S*'s belief that *p* is the result of the noninferential operation of a primal system, and the relevant process is reliable, then the belief that *p* is prima facie justified.

(Lyons 2009, p. 177)

Note that a "primal" system is simply a system that is inferentially opaque and has resulted from learning and/or innate developmental processes. In effect, Lyons proposes that reliability is only sufficient for the prima facie justification of *basic* beliefs. Let's now review how this might help in solving some of the classic problems for process reliabilism.

5.5.1 Generality and Clairvoyance Revisited

Although it does not provide an outright solution to the Generality Problem, inferentialist reliabilism at least helps to rebut some of the counterexamples that have been raised by Conee and Feldman (1998). Take a look at the following scenario Conee and Feldman present as a problem for some versions of reliabilism:

> [S]uppose that Jones and Smith both respond to the same features of a visual input with the belief that there is an elm tree present. Suppose that this input will occur only when there is an elm tree present—it is a distinctive look of an elm leaf, say, the visual appearance of a particular quantity of tiny notches around its edge. Finally, suppose that Smith knows what she is seeing, while Jones is applying some ridiculous and unjustified sort of numerology to the topic. Jones plucks from thin air the idea that the magic number for elms is nine. Jones gets a nine for the tree whose leaf he beholds by counting the number of those distinctive elm notches along the edge of a leaf, and dividing by six, his "tree number".
>
> (Conee and Feldman 1998, p. 15)

In this scenario, Jones and Smith both have a belief that is the result of a fully reliable process, even though the reliability of Jones' process is a matter of mere luck. Without a principled answer to the question which precise type of process—as there are many—should be reliable to account for justified belief, this example thus presents process reliabilism with a problem. Jones surely is not justified in his belief that there is an elm tree present, but process reliabilism has not provided a principled answer to avoid this conclusion.

With inferentialist reliabilism we can now explain the relevant difference between Smith and Jones, thereby eliminating this counterexample. Smith's belief, but not Jones', results from the non-inferential operation of an inferentially opaque cognitive system that has developed by learning about the distinctive look of elm trees. Smith thus has a basic belief resulting from a reliable process, and is justified. In contrast, Jones' belief results from a line of inferential reasoning in which some of the premise beliefs are unjustified. This makes Jones' belief non-basic, and therefore in need of the satisfaction of other conditions to make it justified. These other conditions include, just as in Goldman's conditions for belief-dependent processes, that the premise beliefs in the line of reasoning are themselves justified (Lyons 2009, p. 178). But since it is clear that Jones' beliefs about numerology are not justified, it is also clear that his belief that there is an elm tree in front of him is also not justified. Thus, this example of the Generality Problem no longer presents the reliabilists with any difficulties.

The distinction between basic and non-basic beliefs, and the requirements for justification that go with it, might also help in solving clairvoyance-type problems. Indeed, Lyons contends that these problems precisely have to do with the fact that reliabilist theories have not correctly distinguished between basic and non-basic beliefs (Lyons 2009, p. 122). The clairvoyant beliefs Norman has are presumably not basic, as we do not have any cognitive systems for clairvoyance that satisfy Lyons' etiological constraints (i.e., that have developed due to innate developmental processes and learning). Similarly, Truetemp's tempucomp also does not count as the right sort of cognitive system to produce basic beliefs. The beliefs of both Norman and Truetemp thus count as non-basic, and need more than mere reliability to be justified.

To support this analysis of clairvoyance-type cases, Lyons provides the following variant of the original clairvoyance case in which the posited cognitive system is relevantly different from Norman's:

> **Nyrmoon:** Nyrmoon is a member of an alien species for whom clairvoyance is a normal cognitive capacity, which develops in much the same way as vision does for humans. Members of Nyrmoon's species have specialized internal organs that are receptive to the highly attenuated energy signals from distant events; as an infant, all was a "blooming buzzing confusion" for Nyrmoon, until, like everyone else, he learned to attend selectively, recognize various objects, and filter out coherent distant events. Nyrmoon, however, is so extremely unreflective that he has no beliefs (a fortiori, no justified beliefs) about the reliability of his clairvoyance. One day he forms, as the result of clairvoyance, the belief that his house is on fire (which it is).
>
> (Lyons 2009, p. 119)

According to Lyons, Nyrmoon is, unlike Norman, perfectly justified in his belief that his house is on fire. The relevant difference, claims Lyons, is that Nyrmoon has a specialized clairvoyance system that developed in the right way, whereas Norman merely happens to reliably acquire clairvoyant beliefs.

Now even though I agree with Lyons that Nyrmoon is intuitively justified, I think this example does little to support the idea that this intuition arises because of the presence of a certain type of cognitive system. Nyrmoon's case is set up in such a way that far too many parameters have changed from the case of Norman, as Peter Graham (2011) also notices. For instance, not only does Nyrmoon now have a clairvoyant system that developed in the right way, he also is a member of a species that has such systems, which makes it likely that the system has been selected for by evolution. Moreover, Nyrmoon has also used his clairvoyant system before, and has apparently learned to apply it through trial and error, which means that he has some knowledge (if only implicitly) about when to trust the system's outputs. These other factors could just as well be mentioned as an explanation of the relevant difference between Nyrmoon and Norman. Indeed, even internalists could appeal to the latter factor to explain the difference between Nyrmoon and Norman, and it's clear that they would not accept anything of the sort proposed by inferentialist reliabilism.

Still, Lyons' explanation of the cases of Norman and Truetemp *might* be correct, it's just that it has not been sufficiently motivated. But what about the case of Bill and his blindsight? Recall that Bill is an imagined blindsighter who still forms beliefs about what goes on in the blind part of his visual field even though he has no idea about his own reliability. Although his visual cortex is damaged, this need not mean that the way Bill's beliefs are now formed is relevantly different from the way they were formed before he lost experience in his blind spot. The cognitive system that forms the visual beliefs might still be working in the same way, even though, in actual blindsighters, *another* cognitive system inhibits these beliefs from being formed. In that case, the Blindsight case would remain problematic for Lyons' inferentialist reliabilism.

However, one could just as easily suppose (especially because the case is merely hypothetical) that Bill's cognitive system has changed due to external factors that have nothing to do with learning or innate developmental processes, thereby again violating Lyons' etiological constraint on cognitive systems that are capable of providing basic beliefs. This would also preclude Bill's belief from being justified.

Lyons thus appears to have found a way in which reliabilism can answer the problems raised by clairvoyance-style cases. However, in the next section I will argue that these answers are unfortunately entirely dependent on a restriction in Lyons' definition of basic beliefs that is not well motivated. This means that inferentialist reliabilism will still lack a *principled* answer to these cases.

5.5.2 *The Etiological Constraint*

Let's briefly recapitulate where we are. Classic process reliabilism faced several counterexamples to its thesis that reliability is necessary and sufficient for prima facie justified beliefs. The New Evil Demon scenario attempts to show that reliability is not necessary for justification, while clairvoyance-style cases attempt to show that it is not sufficient for justification. Lyons' inferentialist reliabilism is specifically meant to address the second kind of counterexamples by first, distinguishing between basic and non-basic beliefs, and second, maintaining that clairvoyance-style cases are about non-basic beliefs. Given that inferentialist reliabilism holds that reliability is *not* sufficient for the justification of non-basic beliefs, this prevents the unwanted consequence that clairvoyance-type scenarios lead to justified beliefs.

To answer clairvoyance-style problems, inferentialist reliabilism thus makes the important strategic move of distinguishing between basic and non-basic beliefs. One might think that the whole idea of distinguishing between basic and non-basic beliefs is ad hoc in a reliabilist framework. Process reliabilism is a theory that claims that the justification of beliefs has to do with the reliability of the processes that produced them, so why distinguish between two classes of beliefs that have different conditions for justification? If reliability is sufficient in the one case, then why not in the other? However, even a reliabilist can acknowledge that there are relevant epistemic differences between two classes of beliefs. Some beliefs are based on other beliefs, while others are not. This just comes down to an acceptance of foundationalism, and every foundationalist will have to admit that these classes of beliefs are relevantly different. So there is nothing ad hoc in distinguishing between two classes of beliefs that each have different conditions for justification, even in a reliabilist framework.

However, there is another aspect to Lyons' theory that really is more ad hoc, or at least, not well motivated. This aspect has to do with the proposed conditions for being an epistemically basic belief. Lyons' approach is to start with the intuition that perceptual beliefs are basic, then to supply a definition for perceptual beliefs, and finally to generalize this definition to encompass all basic beliefs. I share the intuition that perceptual beliefs are basic, and I also think that Lyons' definition of them, i.e., that "a belief is a perceptual belief just in case it is the output of a perceptual system" (Lyons 2009, p. 87), is at least on the right track. Looking to cognitive science to define perceptual systems and thereby perceptual beliefs surely is a good way to find out which of our actual beliefs should count as perceptual. However, when this *psychological* definition is used as a starting point from which to generalize to a definition of basic belief, one should be careful to subtract all parts that are not *epistemically* relevant.

It is at this point that problems arise for Lyons' theory. Recall Lyons' definition of basic belief:

> [...] [B]asic beliefs are those that are the outputs of a certain kind of operation of a certain kind of cognitive system, in particular, an inferentially opaque cognitive system, which results from learning and/or innate developmental processes, where the system is not basing its outputs on any doxastic inputs.
>
> (Lyons 2009, p. 166)

It certainly appears to be epistemically relevant that the cognitive system in question is inferentially opaque. Just suppose, in contrast with this condition, that I had access to the fact that some of my beliefs were used by my perceptual system in producing its outputs. Specifically, suppose that I could introspectively ascertain that Jack's looking angry was caused by my prior belief that Jack is angry.[13] In that case it seems that my belief that Jack looks angry would not be epistemically basic, because I would be aware of an inferential basis for the belief. Moreover, it also seems that my belief that Jack looks angry would become unjustified in this scenario, because my knowledge of the causal etiology of my experience would give me reason to be skeptical of its trustworthiness (given that it would be the same whether or not Jack actually looks angry). So there are good reasons to take inferential opacity as an epistemically relevant condition for being a basic belief.

Similarly, the requirement of the cognitive system's not basing its outputs on any doxastic inputs is also epistemically relevant. Even if I am not aware of the fact that a certain cognitive system bases its outputs on doxastic inputs, the mere fact that it does is already epistemically important. Just suppose that I am implicitly inferring that it is likely that you have a gun because of a certain racially prejudiced belief of which I am not even aware. The belief that you are likely to have a gun will be unjustified because of the fact that it is based on the unjustified racially prejudiced belief, whether or not I am aware of it. This goes back to the idea that a belief can be evidentially justified only if its evidential justifier is itself justified.

However, it is not so clear what is epistemically (rather than psychologically) relevant about Lyons' etiological condition that the cognitive system supplying basic beliefs should be the result from learning and/or innate developmental processes. Although Lyons tries to support his etiological constraint with the case of Nyrmoon, I have already argued why that case is insufficient for its purpose in the previous section. This makes the etiological constraint into an undermotivated condition in Lyons' definition of basic belief. Given that clairvoyance-type problems are excluded purely on the basis of the etiological constraint, this is a major problem for the theory.[14]

To make this point more clear, let's look at the Truetemp case again. According to Lyons, Truetemp is not justified in his belief that it is 104 °F, because his tempucomp has not resulted from learning and/or innate developmental processes, thereby making the beliefs it produces non-basic (Lyons 2009, p. 145). Although this judgment about the case is in line with our intuitions, Lyons' theory only provides it because it uses the undermotivated etiological constraint on systems capable of producing basic beliefs. Importantly though, the example can be adapted in a way that intuitively makes Truetemp justified in his belief, even though the cognitive system producing it still does not count as the right sort of cognitive system.

---

[13]The example is from Susanna Siegel (2012).

[14]Lyons is not adverse to dropping the etiological constraint if it turns out to be unprincipled, and suggests that our intuitions about clairvoyance-type cases might be mistaken (Lyons 2009, p. 164–5). I think it should be recognized that the etiological constraint *is* unprincipled, but I also think there is another way to accommodate our intuitions about clairvoyance-type cases. I will present this proposal in the next chapter.

Truetemp*: Suppose that, instead of merely having a tempucomp implanted in his brain, Mr. Truetemp's brain is also rewired in such a way that it is *as if* he has had this new sensory modality for quite some time, and as if he had fully adapted to it through learning. Truetemp's new sense modality is fully integrated with his other senses, and Truetemp trusts in the output of his tempucomp just as he trusts in the output of his other senses. In fact, the surgery is such a success that Truetemp does not even notice any relevant difference between his tempucomp and his other senses.[15]

In this revised scenario, Truetemp appears to be justified in his belief that it is 104 °F even though this belief still isn't the result of the noninferential operation of an inferentially opaque system that developed in the right way. This shows the etiological constraint to be epistemically irrelevant.

One might try to accommodate the Truetemp* scenario in Lyons' theory by claiming that it is unclear what it means to rewire the brain in such a way that it is as if Truetemp had fully adapted to his new sensory modality through learning. Maybe this just means that Truetemp now also believes that his ability to measure the temperature is reliable, or believes that he should trust in his temperature-measuring abilities. If this is true, then his belief that it is 104 °F could be taken to be based on those beliefs, thereby making it into a non-basic, but inferentially justified, belief.

Unfortunately, there are two problems with this response to Truetemp*. First, the beliefs that are the supposed consequences of Truetemp's brain-rewiring do not appear to be justified themselves, as they are implanted in Truetemp's brain instead of being the output of the right kind of cognitive system. And if the premise beliefs in a conditionally reliable inference process are not justified themselves, then the concluding belief also cannot be justified. Second, if Truetemp's beliefs about his tempucomp-ability would make his temperature-beliefs inferential, then *all* our current perceptual beliefs would be inferential as well, since we surely believe that our senses are reliable and trustworthy. And this is exactly what Lyons wants to deny. Even if it is true that we *could* base our perceptual beliefs on our higher-order beliefs about the reliability of our own perceptual capacities, this does not mean that we actually do base them on these considerations. And according to the reliabilist framework, the *actual* belief-forming process should be considered to determine whether a belief is (prima facie) justified.

At any rate, relying on higher-order beliefs to accommodate Truetemp* does not solve the bigger issue of the undermotivated nature of the etiological constraint. Think, once again, of cognitively unsophisticated cognizers. Suppose that a creature only capable of first-order belief developed a tempucomp-like sense because of exposure to radiation. It had always been capable of feeling heat, but could now discriminate temperatures far more precisely. On the basis of this new sense it forms the belief that its favorite food is located at places with precisely this temperature, and, because of this, even manages to locate more food. The intuitive pull to judge

[15] James Beebe (2004) uses this variation on the Truetemp case to show that the original case is actually underdescribed and therefore leads to the mistaken conclusion that Truetemp is unjustified in his beliefs. I think this response focuses too much on the contingencies of human belief acquisition to work as a general response against Truetemp-style cases.

this creature to be unjustified in its beliefs appears to be at least less strong than in the original Truetemp-case, if it is not absent completely. This again puts pressure on the epistemic relevance of Lyons' etiological constraint, given that this creature's cognitive system does not satisfy it.

However, without the etiological constraint, inferentialist reliabilism will lack a principled way of accommodating the examples of clairvoyance, Truetemp, and blindsight. What's more, inferentialist reliabilism has no immediate bearing on the New Evil Demon Problem. So there seems to be sufficient reason to look at another externalist way of developing process reliabilism.

5.6 Proper Functionalism

In the previous section we saw that inferentialist reliabilism made use of an etiological constraint that was not well motivated from its own reliabilist perspective. In this section we'll look at another way of developing process reliabilism that does create a motivation for allowing etiological constraints into one's theory of justification: *proper functionalism.*

Proper functionalist theories of justification, or more precisely, warrant, have been defended by e.g., Alvin Plantinga (1993) and Peter Graham (2012, 2014).[16] Such theories stress the importance of a cognitive system's functioning *properly* rather than its functioning *reliably* in relation to providing epistemic justification. However, reliability is certainly still an important factor in the proper functionalist framework.

Where classic process reliabilism focuses on the de facto reliability of a certain cognitive process, proper functionalism focuses instead on whether a cognitive system is functioning according to its *aim* of reliably producing true beliefs. And it is precisely because proper functionalism evaluates cognitive systems or processes with regard to their aim rather than their actual outcome that it requires that certain kinds of etiological conditions be fulfilled before there can be any kind of epistemic justification. Without the right kind of etiology, there simply will be no aim or function to whatever causal process is responsible for producing a subject's beliefs.

So what is the right kind of etiology? According to Plantinga (1993), the right kind of etiology has to do with a cognitive system's having a certain *design plan* that is successfully aimed at the truth. Although intelligent design might be the first thing

[16]To be precise, Plantinga defends a view about warrant (the property that added to truth is sufficient for knowledge) and Graham a view about entitlement (which can be contrasted with the warrant one gets from reasons, called 'justification' in his terminology). I will simply discuss their views in terms of justification for ease of exposition. The points against these theories will still stand when translated to points about warrant and entitlement.

that springs to mind, the theory might work just as well by using natural selection as the relevant 'designer'. Even so, using the notion of a design plan is problematic for two reasons.

First, having a design plan that is successfully aimed at the truth is not sufficient for producing justified beliefs, as the case of Truetemp shows. Truetemp's tempucomp has been designed to reliably produce true beliefs about the temperature, yet Truetemp still intuitively lacks justified beliefs.[17]

Second, having a design plan that is successfully aimed at the truth also appears unnecessary for justification. Take Sosa's use of the case of Swampman as an illustration of this point (Sosa 1993). Swampman comes into existence as a normal human subject's physical duplicate by a random event involving swamp gas, a dead tree, and a lightning strike. Swampman has no design plan—he is the result of a random accident—but acts precisely as a normal human being would act. Now one might be disinclined to count Swampman as a real human being, given that he was not the offspring of a human father and mother. One might also hesitate in immediately granting Swampman the full range of contentful beliefs, given that there is no answer to such questions as whether Swampman refers to water (H_2O) or twin-earth water (XYZ) when he thinks "I would like to have some water" (Putnam 1973). Nevertheless, after Swampman has been sufficiently embedded in his environment by, e.g., coming into contact with water, there seems to be a strong pull to assign at least some justified beliefs to Swampman. But on Plantinga's proper functionalism, we should deny that Swampman can have *any* justified beliefs.

Even if one takes the example of Swampman to be problematic, perhaps because it is too far removed from actuality, then one can make the same point by using actual cases in which people have received new perceptual faculties (Beebe 2004). Even though these perceptual faculties were no part of a human being's original design plan, after people have learned to sense with them they should certainly be able to produce justified beliefs in a manner similar to that of other perceptual faculties.

So instead of focusing on design plans, it seems that we should focus on a rather different type of etiology to explain where the proper functions necessary for justification arise from. According to Graham, the right kind of etiology is a *consequence etiology* that "*explain*[s] why something exists or continues to exist in terms of its consequences, because of a feedback mechanism that takes consequences as input and causes or sustains the item as output" (Graham 2014, p. 18). Now, one example of such a consequence etiology comes from natural selection. In the envisaged case of natural selection, a system produces a result that is beneficial to the members of a certain species, thereby enabling these members

[17]Although Plantinga (1996, p. 333) attempts to explain this case as one in which Truetemp's beliefs are *defeated*, his specific explanation appeals to beliefs that are not really up to the job of defeat. For instance, Plantinga mentions that Truetemp might have a defeater in his belief that he is constructed like other people who all lack his precise temperature-reading ability. But surely all people with extraordinary belief-forming abilities have these kinds of beliefs, and we shouldn't suppose that all these abilities are thereby incapable of leading to justified beliefs (think of, e.g., brilliant mathematicians or people with extraordinary eyesight).

to reproduce. In this way, the systems get replicated in the descendants with the function of again producing the same beneficial thing(s). Note that a system can have several functions simultaneously: many systems will have contributing to survival as one of their functions, but the way in which they contribute to survival (e.g., by pumping blood in the case of the heart) can also be one of its functions (Graham 2012, p. 474).

But natural selection is not the only way in which a consequence etiology can arise. If this were true, then it seems that we could still never acquire *new* ways of gaining justified beliefs, and the cases of Swampman and new perceptual faculties would remain as problematic as before. Fortunately, other examples of consequence etiologies are easily found. For instance, take the consequence etiology that is at work in cases of *trial-and-error-learning*: a certain behavior produces a beneficial result, and so the subject gets reinforced in using that same behavior again with the function of getting the beneficial result again (Graham 2014, p. 31). And the way in which new perceptual faculties are sustained also fits a consequence etiology: after adaptation and learning, these faculties reliably produce true beliefs and are continually used because they reliably produce these beliefs. Hence, such new perceptual faculties will be able to produce justified beliefs after they have been successfully used before.

The proper functionalist framework thus has a way of upholding an etiological condition on the types of cognitive systems capable of producing justified beliefs. A justified belief is the output of a properly functioning cognitive system which aims at reliably producing true beliefs. Only systems that have developed from innate conditions or learning are thus systems capable of producing justified beliefs. We will now turn to see how this proper functionalism plays out exactly with regard to the Clairvoyance Problem and the New Evil Demon Problem.

5.6.1 Clairvoyance and New Evil Demon Revisited

Given the way in which Lyons' etiological constraint worked in cases of clairvoyance, it should not be too difficult to see how the proper functionalist theory is able to address these cases as well (Graham 2012). The difference is that proper functionalism is not committed to a distinction between basic and non-basic beliefs, although it would perhaps be well advised to also take this distinction into account. Anyway, returning again to the case of Norman, since Norman's randomly acquired clairvoyant faculty does not (yet) have a proper function, it is also not (yet) capable of producing justified beliefs. There is simply no consequence etiology for the clairvoyant capacity that explains why this capacity exists on the basis of its consequences. However, if Norman were to derive beneficial results from the fact that his newly acquired cognitive faculty reliably produced true beliefs—e.g., if he remained alive because the system kept him from entering into potentially dangerous situations—then it would acquire the proper function of reliably producing true beliefs, and thereby would acquire the capacity to produce justified beliefs.

For Truetemp, the case is slightly different. Although the existence of the tempucomp can be explained in terms of its reliably producing true beliefs (after all, that is what the mad scientists have designed the thing to do), it's *acquisition* by Truetemp cannot be explained by its epistemic consequences (Graham 2012, p. 478, n. 14). Here it becomes clear that it is not just the existence of a cognitive faculty that needs to be explained by its consequences, a subject's acquisition of the faculty should also be explained by these consequences. That is, the history of the acquisition should connect to the history of the faculty in the right way. For instance, Truetemp's acquisition of his visual system *is* explained by the consequences of that system: because this system reliably produced true beliefs in his ancestors, they were able to pass on this system to Truetemp himself. But the way in which the tempucomp was implanted in Truetemp is not connected to the epistemic consequences of the tempucomp in a similar manner.

Finally, focusing now on the case of blindsight, although Bill's visual system does have a proper function, the system itself is not functioning properly; it's not functioning in the way it was when it got selected in Bill's ancestors. Now, once again, if Bill continues to use his new blindsighting ability, then it would probably become capable of producing justified beliefs. But note that in such a case Bill will probably also know about the reliability of his own blindsight, thereby making the case an uncontroversial case of acquiring justified beliefs.

So proper functionalism, in contrast to inferentialist reliabilism, does appear to have a motivated response to several of the clairvoyance-style cases discussed above. But there's more to proper functionalism than just that. Because it no longer focuses on the de facto reliability of one's cognitive processes, proper functionalism is also able to provide a motivated response to the New Evil Demon Problem (Graham 2012). Recall Ned, our luckless demon-deceived subject, who forms beliefs in accordance with his demon-induced experiences. Even though the cognitive process he uses is de facto unreliable, all of his cognitive systems are still functioning *properly*. Just as a car on the lift can pass its inspection with flying colors even though it will not actually drive you anywhere, so can Ned's perceptual system be working properly without it actually reliably producing true beliefs. Ned's cognitive systems are functioning in the way they were functioning when they were naturally selected for their beneficial effects, and so they are performing exactly as they should perform given their consequence etiology. This means that, according to proper functionalism, Ned's beliefs are fully justified: they are the output of a specific cognitive system that is functioning in the way it was designed to function. It's just that, under these conditions, functioning in this way does *not* reliably lead to true beliefs. Fortunately, this no longer is a requirement for justification once one accepts the proper functionalist's view.

So proper functionalism has principled answers to both the Clairvoyance Problem and the New Evil Demon Problem. This means that it surely is a theory that deserves to be taken seriously. However, in the following section we'll see that even proper functionalism has some important difficulties left.

5.6.2 Problems with Proper Function

Even though proper functionalism is off to a good start with its responses to the Clairvoyance and New Evil Demon Problem, it's not home free yet. First, there is still a question with regard to the *necessity* of proper function to acquire justified beliefs. Although Graham's (2014) notion of having a consequence etiology allows that the earlier mentioned Swampman acquires justified beliefs after a certain amount of time (when his cognitive faculties have acquired the proper function of reliably producing true beliefs), it still does not allow Swampman to acquire justified beliefs *immediately*.[18]

Now one might question whether our intuitions are really a good enough guide to these fine-grained differences, but there is a theoretical question at the bottom of the illustrative case of Swampman: why can only cognitive systems that have proper functions produce justified beliefs? Graham's response to this question is that it only makes sense to think that a cognitive system has satisfied a certain epistemic norm if that cognitive system also has a proper function on the basis of which one can decide whether the system has functioned in the way it should (Graham 2012, pp. 467–8). In the absence of proper function, there is no way a system should behave, and so no sense can be made of the idea that the system has performed in the epistemically right way. And beliefs can only be justified if they are the result of a cognitive process that has performed in the way it epistemically should have.

But one can question whether epistemic norms can only arise from these kinds of teleological considerations. We might also simply apply epistemic norms that *we* derive from things we value epistemically: e.g., we value truth, so we evaluate belief-forming processes that reliably produce true beliefs positively. This might be sufficient to provide an epistemic norm for such belief-forming processes without appealing to the teleological considerations that Graham favors. The difference would be that the epistemic norms are now no longer fixed by the history of the cognitive systems themselves, but are rather applied to those cognitive systems by us, the people who care about systems that reliably produce true beliefs. Perhaps one might worry that this makes the property of justification 'unnatural' or 'constructed', but such worries are not really grounded when one carefully considers what is going on. Even if the epistemic norms for cognitive systems derive from things we value as epistemic agents instead of coming from the cognitive systems themselves, this does not make the properties having to do with these norms any less real.

So Graham's account of naturalizing epistemic norms is not the only way one can go, and the alternative could allow there to be justification even in the absence of proper function. Still, it's difficult to make this objection stick without finding illustrative cases where proper functionalism *clearly* provides the wrong verdict. The case of Truetemp* (where Truetemp not only gets a new tempucomp but also a

[18]Also see Simion (2016) for more difficulties with spelling out the exact requirements for Swampman's having the right kind of consequence etiology.

rewiring of his brain so that it is as if he has been using this tempucomp for a long time) and the case of the radiation-exposed animal with a tempucomp-like sense might go some way in that direction (see Sect. 5.5.2). But there is also a variation on the New Evil Demon scenario that could be used to make a similar point.

In the original scenario, Ned supposedly is a human being who at some point in his life comes to be radically deceived by an evil demon that produces non-veridical perceptual experiences of his environment. Given that Ned's perceptual system is still functioning properly though, the beliefs he forms on the basis of these experiences still come out as being justified according to proper functionalism. However, a problem appears if one reverses this standard New Evil Demon scenario to create a Reversed Evil Demon scenario. Suppose that Red is a subject who has been deceived by an evil demon from his birth on, and just now has been released from this demon's grasp into the normal world again—without him noticing of course. Given that Red's cognitive systems have not been sustained by a consequence etiology that figured reliably produced true beliefs, the proper function of these systems will no longer be the same as ours. They might even have gained reliably producing *false* beliefs as their function, as the demon has attempted to produce as many of these beliefs as possible and has kept Red alive for just this purpose. This means that when Red is released from the demon's grasp into the normal world, he will not be immediately able to form justified beliefs, as his cognitive systems will not be functioning properly with the goal of reliably producing true beliefs. Yet surely it is plausible that once Red is released into the normal world again, he is immediately able to form justified beliefs. His sensory systems are reliably picking up on all of the information in his environment, and he has no reason to doubt anything his senses tell him. This again points to the fact that proper function is not necessary for justification, and that the proper functionalist story cannot be fully right as it stands.

Next to this problem about the necessity of proper function for justification, one can also use a variation of the clairvoyance case to cast doubt on the *sufficiency* of proper function for justification. Take the following variation on the cases of Norman and Nyrmoon:

> **Norbert:** Norbert is the son of a mother and father who both have reliable clairvoyant abilities and have been able to reproduce because of the benefit these clairvoyant abilities have provided for them. However, there are not (yet) many people who have these clairvoyant abilities, and their existence is kept secret. The abilities are due to specialized internal organs that pick up on special energy waves in the environment, and then output brief visual images which represent that such-and-so is currently the case at some distant place. Furthermore, the abilities usually become active quite suddenly some time after puberty. Norbert's parents have decided not to tell him about the existence of his clairvoyant abilities, and Norbert has no evidence for or against their existence in general or his own having them. Some time after puberty, Norbert suddenly experiences a brief visual image of the President being in New York City and on that basis believes that the president is in New York City.

Norbert's clairvoyant faculty has the right kind of consequence etiology, because the acquisition of this faculty is explained by the fact that it has provided a certain benefit to Norbert's ancestors. Furthermore, the clairvoyant faculty is reliable, it has developed from innate conditions and is also inferentially opaque. It thus seems to have everything a proper functionalist could want from it, yet the belief it produces still appears to be unjustified for exactly the same reason as before: from Norbert's own subjective perspective there seems to be no reason to trust the visual image that briefly flashes before his eyes. This means that there is reason to doubt that proper function is fully sufficient for justification. At the very least, even if one were to assume that a proper functionalist story is correct in general, its *diagnosis* of the Clairvoyance Problem remains problematic.

Even though proper functionalism thus has principled responses to the original New Evil Demon and Clairvoyance problem, variations of these cases still put some pressure on the proper functionalist theory. The next chapter will therefore provide an alternative way of dealing with these problems that makes use of some insights derived from the epistemological disjunctivist theory of perceptual justification.

5.7 Conclusion

In this chapter I have looked at the prospects of a non-evidential theory of perceptual justification, namely process reliabilism. Even in its early version promoted by Goldman (1979), process reliabilism is presented as a foundationalist theory of justification according to which, for a specific class of beliefs (viz., basic beliefs), the reliability of the relevant belief-forming process is necessary and sufficient for prima facie justification. This distinction between basic and non-basic beliefs is emphasized and refined in Lyons' (2009) inferentialist reliabilism, where basic beliefs are taken to be those that follow from the non-inferential operation of an inferentially opaque cognitive system that developed in the right way.

Although Conee and Feldman (1985, 1998, 2002) pose a challenge for process reliabilism with their Generality Problem, it does not appear to be a knock-down objection to the theory. There is no reason to assume that there are many different functions actually operative in the way cognitive systems map perceptual input to belief output. The Generality Problem does raise a challenge for reliabilists to explain how we attribute justification to others without having a clear idea of the precise cognitive processes whose reliability supposedly determines that justification.

A bigger problem for process reliabilism is presented by the classic counterexamples against the thesis that reliability is both necessary and sufficient for prima facie justification. The New Evil Demon scenario puts pressure on the necessity claim: the perceptual beliefs of subjects under the influence of an evil demon are not the result of a reliable process, but they still appear to be justified. The standard reliabilist response that the relevant sort of reliability, or our judgments about that reliability, might be relative to the actual world simply does not appear to be very plausible without further development.

Clairvoyance-style counterexamples attack the sufficiency claim of process reliabilism. Norman the clairvoyant, Mr. Truetemp, and Bill the blindsighter all appear to lack justified beliefs, even if these beliefs result from perfectly reliable cognitive processes. Lyons attempts to accommodate this intuition by pointing to the non-basicality of the beliefs under consideration: all these beliefs do not stem from a cognitive system that resulted from learning and/or innate developmental processes, and therefore require more than mere reliability for their justification. I have argued that this response to the examples is not well motivated: the etiological constraint Lyons puts on cognitive systems capable of producing basic beliefs is not clearly epistemically relevant and so should not be used to discount beliefs as basic.

In contrast, proper functionalism provides a development of process reliabilism that does give a principled reason for looking at the etiology of our cognitive systems. According to proper functionalism, what matters for justification is that one's beliefs are produced by a properly functioning cognitive system that has the aim of reliably producing true beliefs. Such aims and proper functions only make sense in the light of a system's precise etiology, thus providing an important reason for requiring etiologies of a specific type. This way of cashing out justification gives proper functionalism the tools to respond to both the Clairvoyance and the New Evil Demon Problem. Norman's clairvoyant faculty lacks proper function, hence Norman lacks justification for his clairvoyant beliefs; Ned's perceptual faculty has a proper function and functions according to it, hence Ned has justified perceptual beliefs. Unfortunately though, one can produce variants of the Clairvoyance and New Evil Demon problem which challenge the thesis that proper function is necessary and sufficient for having justified beliefs.

This means that more works needs to be done to make process reliabilism into a viable theory of perceptual justification. In the next chapter, I will attempt to do just that by providing more detailed answers to both the Clairvoyance and New Evil Demon Problem. These answers will make use of the epistemological disjunctivist insight that we have higher-order capacities at our disposal that enable us to acquire specific types of reasons for and against our own first-order beliefs. This insight can be used in a straightforward way to help accommodate clairvoyance-style cases, and is also helpful in an explanation of our intuitions about the New Evil Demon scenario.

Chapter 6
A Higher-Order Rejoinder for Reliabilism

6.1 Introduction

In the last chapter we saw that process reliabilism was confronted with several problems, at the foremost of which were the lack of a good answer to the New Evil Demon Problem and the Clairvoyance Problem. If reliability is indeed necessary and sufficient for the justification of a specific class of beliefs, then subjects misled by an evil demon are unjustified in their perceptual beliefs, while subjects forming beliefs on the basis of an extravagant but possibly reliable process, such as blindsight or clairvoyance, are justified in their beliefs. Given that we have the intuition that this consequence is false, it seems that process reliabilism must also be false. Process reliabilism is thus in need of a good way to accommodate the New Evil Demon Intuition and the Blindsight (and Clairvoyance) Intuition.

In this chapter I discuss what I take to be the best way for reliabilism to account for the Blindsight Intuition and to solve the Clairvoyance Problem. This account starts from some of the insights of previous chapters, and focuses especially on the importance of higher-order capacities and their impact on first-order processes. Given the problems of experientialist and epistemological disjunctivist theories of perceptual justification, this will leave process reliabilism as one of the better theories of perceptual justification.

In Sect. 6.2, I briefly rehearse the way in which epistemological disjunctivists could respond to the Blindsight Intuition, i.e., by appealing to the absence of factive reasons. I then go on to show how a process reliabilist could make use of a similar response by accommodating the possibility of higher-order evidential reasons that one sees that *p*. To make this account of higher-order reasons work, though, the process reliabilists needs to show that these introspective beliefs can be justified in a process reliabilist way. I discuss the possibility of such a process reliabilist theory of introspective justification, especially in relation to mental state *attitude* types, in Sect. 6.3.

H. Ghijsen, *The Puzzle of Perceptual Justification*, Synthese Library 377,
DOI 10.1007/978-3-319-30500-4_6

In Sect. 6.4, I then go on to show how a process reliabilist response to the problems of Blindsight and Clairvoyance could be strengthened by appealing to higher-order *defeat*. It's plausible that our introspective mechanisms not only have the function of outputting beliefs about our own mental states, but also the function of monitoring and controlling the first-order processes which lead to those beliefs. I argue that a proper functionalist account of defeat could utilize these monitoring mechanisms to explain why clairvoyants and blindsighters have unjustified beliefs, and tentatively suggest a way in which this could be incorporated in a hybrid theory of proper functionalism and process reliabilism, or, with some amendments, into a full-blown process reliabilism.

In Sect. 6.5, I then revisit the New Evil Demon Intuition. I first show how a hybrid theory of justification could accommodate the intuition, and then also provide a way of explaining the intuition by pointing to a natural experientialist assumption. It's natural that we would take experiences to be our evidence for our perceptual beliefs, given their actual reliability and our ways of justifying our beliefs. However, if my arguments so far have been successful then we should seriously consider rejecting this experientialist assumption.

Section 6.6 then briefly summarizes the crucial ingredients of the proposed non-evidential theory of justification, and discusses a few objections from an experientialist and epistemological disjunctivist standpoint. The final Sect. 6.7, briefly sums up the results of this last chapter.

6.2 Incorporating an Epistemological Disjunctivist Insight

In the previous chapter we saw that none of the discussed varieties of process reliabilism had a fully satisfying answer to the Clairvoyance Problem. Inferentialist reliabilism had to make use of an undermotivated etiological constraint on the type of system capable of producing basic beliefs, and even a full-fledged proper functionalism was susceptible to a new variant of the original clairvoyance case. However, earlier we noticed that epistemological disjunctivism had a way of accommodating the Blindsight Intuition that could be used independently of the precise epistemological disjunctivist theory. This explanation might thus also be exactly what the process reliabilist needs to answer the Clairvoyance Problem.

According to epistemological disjunctivism, our blindsighter Bill lacks perceptual justification for his belief that *p* because he lacks access to a factive reason: even though Bill has a reliable perceptual process that outputs the belief that *p*, Bill still does not have any access to the fact that he *sees* that *p*. Now, we've already encountered a good reason not to hold that access to factive reasons is necessary for perceptual justification: this runs into the problem of hyper-intellectualization. Nevertheless, one could still hold that factive reasons are relevant to justification in a different way, a way that can in fact be used to develop an answer to the Clairvoyance Problem from a process reliabilist perspective.

What we need to develop this answer is the distinction between basic and non-basic beliefs, and the distinction between evidential and non-evidential justification. Most process reliabilists will want to allow for the possibility of evidential justification, where a subject's belief is supported and justified by other beliefs rather than its being justified by being the output of a belief-independent belief-forming process. The usual way in which reliabilists account for this type of justification is by appealing to at least conditional justification of the relevant evidential inference, where there is a high truth-ratio given that the input beliefs are true. Now, one instance of an evidential process that is surely going to be conditionally reliable is the one going from the belief that I see that *p* to the conclusion that *p*. If the belief that I see that *p* is true, then the conclusion that *p* is always going to be true as well, given that seeing is factive. So the justificatory force of factive reasons can easily be accommodated by process reliabilists.

However, next to conditional justification, process reliabilists will also want to require at least that the input beliefs to the conditionally reliable inferential process are justified. This is where the distinction between basic and non-basic beliefs becomes important. If the belief that I see that *p* is non-basic, then it stands in need of further evidential support, and we might be off in an evidential regress. If, however, the belief that I see that *p* is basic, then it can be justified without further evidential support, and thereby justify perceptual beliefs immediately.

It would thus be ideal if process reliabilists were able to accommodate the basicality of higher-order introspective beliefs like the belief that I see that *p*. In fact, not much stands in the way of accommodating just that. Recall that on Millar's (2010) account of recognitional capacities, one had lower-order and higher-order recognitional capacities by means of which one could recognize respectively *p* and that one was seeing that *p*. Crucially, for Millar the exercise of these capacities was a success notion: one could not exercise a capacity to recognize that (one was seeing that) *p*, without its actually being true that (one was seeing that) *p*. However, a process reliabilist might well weaken this idea to the following thought. Just as one has reliable perceptual processes that deliver beliefs about the world, so one also has reliable introspective processes that deliver beliefs about oneself. The reliability of these introspective processes provides their output beliefs with justification irrespective of whether the output is actually true. If these introspective processes further satisfy some of Lyons' (2009) intuitive constraints on basicality, i.e., that they are processes of inferentially opaque cognitive systems in which the output beliefs are not based on doxastic inputs, then there seems to be no reason against counting introspective beliefs as epistemically basic.

Before going to a full discussion of a process reliabilist account of introspection, let's first see how this account would enable a process reliabilist to respond to the Clairvoyance Problem. In ordinary cases of perception, we have a type of evidential justification at our disposal by means of which our perceptual beliefs are justified: whenever we see that *p*, we also have justified introspective beliefs that we see that *p* and can thereby evidentially infer that *p* is the case. However, this type of justification is absent in both the cases of blindsight and clairvoyance: Bill does not have the higher-order belief that he sees that *p*, nor does Norman have the higher-order belief that he clairvoyantly perceives that *p*.[1] So the process reliabilist can allow that there is a type of justification that is absent in cases of clairvoyance and present in ordinary cases of perception, i.e., that of evidential justification by factive reasons.

Even if one agrees with all of this, then the process reliabilist account might still fall short. For instance, not only do we have the intuition that Norman is unjustified in his clairvoyant belief, we also have the intuition that unsophisticated cognizers are fully justified in their perceptual beliefs. Yet these unsophisticated cognizers all lack higher-order beliefs that they perceive that *p*, and so lack a type of justification that is supposedly relevant to our intuitions. So what explains why we have different intuitions in these cases, even though in both cases the agents lack the relevant type of evidential justification? Perhaps one might claim that we implicitly raise our standards for justification once we discuss more sophisticated agents like ourselves. But I think there is a more promising solution when one considers the kind of metacognitive processes that we are able to engage in as sophisticated cognizers. Sometimes our first-order beliefs are simply defeated by our higher-order processes, thereby leaving clairvoyants and blindsighters without justified beliefs even if these beliefs were produced by a reliable process. Before going to the precise details of this account, though, it would be good to consider whether reliabilism really can make sense of the way in which introspective beliefs are justified. Without a good account of our higher-order introspective beliefs and their justification, this response simply will not even get off the ground. That is why the next section will now turn to discussing just such an account.

6.3 Process Reliabilism About Introspection

Like Millar (2010), I take it to be relatively uncontroversial that we usually know that we see that *p* whenever we see that *p*. Similarly, we also usually know that we imagine that *p* when we imagine that *p*, or know that we remember that *p* when we remember that *p*, etc. Unlike Millar, however, I also take it that having knowledge that *p* also implies having justified beliefs that *p*. The question is, though, what the source is for this introspective knowledge and justification.

6.3.1 Forcefulness Revisited

It seems that we have already encountered one suggestion to answer this question in Chap. 3. Recall that according to Huemer (2001, p. 77), "even if you have a very vivid, very detailed imagination, or if you have very poor eyesight, you still would

[1] Note that Graham (2012, p. 455) also mentions this type of response to the Clairvoyance Problem.

never confuse seeing a tomato with imagining one." Huemer's explanation for this claim appealed to a certain phenomenological property of perceptual experience: its *forcefulness*. Here's Huemer on the distinction between imagination and perception and its relation to forcefulness:

> An exercise of imagination is distinguished from a perceptual experience chiefly by the fact that the object of the latter seems to the subject to be present then and there, while the object of the former does not seem to exist at all. Again, this has nothing to do with how strong or faint the image is. My "forcefulness" is therefore not equivalent to Hume's "vivacity," insofar as I understand the latter.
>
> (Huemer 2001, p. 78).

Huemer is keen on distinguishing his own proposal from that of Hume because he sees it as problematic. There could be cases of perception that are not very vivid at all, and there could be cases of imagination that are very vivid (after all, someone can be said to have a very vivid imagination). Nevertheless, claims Huemer, no subject looking at a faint photograph will have any temptation to think that he is merely imagining the object on the photograph (2001, p. 79). Now I have already questioned whether no one could ever confuse a perception for an imagination by appealing to the Perky experiments, but here I want to focus on something else. Huemer seems to run together two different questions about perceptual experience, namely:

(1) What phenomenal property is distinctive of perceptual experience?
(2) On the basis of which property do we ourselves distinguish between our perception and imagination?

Not only does Huemer think that there is a distinctive, irreducible phenomenal property of perceptual experience (i.e., forcefulness), he is also committed to the idea that we ourselves distinguish between perception and other experiences on the basis of this property. That is why he emphasizes, against Hume's claim that perception has more vivacity than imagination, that subjects would not confuse perception with imagination on the basis of an absence of vivacity or detailedness.

However, even if one believes that there is a distinctive phenomenal property of perceptual experience, we need not assume that we ourselves distinguish between perception and other mental states on the basis of this property. In fact, once one reflects on what usually goes on (phenomenologically) when one perceives or imagines that *p*, it becomes quite unlikely that we are using phenomenology to identify our perceptions or imaginations for what they are. For instance, it is not normally the case that we first have to examine our own experience before we know it to be an imagination or a perception. We do not ordinarily use the phenomenology of our own experience as *evidence* for the conclusion that we are undergoing an imagination. Similarly, we do not have to examine our imagination to find out what we are imagining. We normally know *that* we are imagining, and *what* we are imagining, without relying on the phenomenology of the imagination itself.

A similar story goes for perception. It certainly seems to be the case that we normally know that we are perceiving when we are perceiving, without relying on the phenomenology of our perceptual experience as evidence for this knowledge—

although one of course could *try* to do this in certain scenarios (e.g., when confronted by a skeptic). But if it is correct that we normally know that we are perceiving and imagining without relying on the perception or imagination itself, then how do we know this, and what justifies this knowledge?

The best answer to this question seems to be externalist in nature, as there is no good empirical reason to assume either that we always have possible access to the cues on the basis of which we form higher-order beliefs about our ongoing mental life, or that the cues we use only consist out of the contents of our mental states. Looking at, for instance, beliefs about our own memory images, it is a plausible empirical hypothesis that we use the *fluency* of our cognitive processing as a cue to assess the trustworthiness of the retrieved memory, rather than the content of that memory image itself (Dokic 2012). What's more, according to the source monitoring paradigm (Johnson et al. 1993), when we have to ascertain the source of our memory images we unconsciously make use of not only the content but also the context of our memory images to determine whether it was, e.g., imagined or perceived. Appealing to such unconscious mechanisms could also help to explain what goes on in the Perky experiment or empirical cases of hallucination: the information that is fed into the cognitive system becomes classified as respectively imaginatory or perceptual, and subjects act in a way that corresponds with this idea.

Once we go down this route of accounting for introspective justification on the basis of reliable unconscious mechanisms rather than an experience's phenomenology, one can even turn Huemer's explanation upside down and account for distinctive phenomenological properties on the basis of the outputs of these unconscious mechanisms. One could still allow that many perceptual experiences have an associated feeling of forcefulness, but one can now explain this feeling as a consequence of an introspective mechanism that outputs beliefs about one's own mental states.

The picture I am suggesting is as follows. We do not distinguish perception from other experiences by using the phenomenology of these experiences as evidence, but instead know the experiences for what they are by means of a reliable cognitive mechanism. Thus, it is ill-suited to look for a special phenomenal feature of perception on the basis of which we know it to be perception: there is no such phenomenal feature. The phenomenal distinctiveness normally associated with perception and imagination might just stem from having the respective experiences in combination with having higher-order beliefs about their nature.

Wouldn't such a proposal fall prey to the phenomenon of known illusion? The phenomenology of some illusions (like the Müller-Lyer illusion) is the same even if one does not believe in the veridicality of one's own experiences. However, I am not suggesting that the subjective distinctness of perceptual experience can be explained only by a higher-order belief that one is currently *veridically* perceiving that such-and-so is the case. Instead, the relevant higher-order belief could be that one is currently 'perceiving' in the sense of gaining information from a certain outside source instead of from an inside source—where gaining information should be understood non-factively. *This* belief can be present even if one does not believe that the current perceptual experience is veridical.[2]

6.3.2 Compatible Accounts of Introspection

According to my reliabilist proposal, a higher-order belief that one is perceiving such-and-so is justified not because of experiential evidence of a certain kind, but because it is the output of a specific sort of reliable cognitive process: an unconscious introspective process. This proposal about introspective justification provides an empirically plausible account of how we distinguish between different mental states (such as perception and imagination), is in line with the immediacy of our introspective phenomenology, and parallels the proposal about perceptual justification.

This type of introspective account might resemble monitoring accounts of introspection that have been defended by, e.g., Nichols and Stich (2003) and Goldman (2006). According to Nichols and Stich, our monitoring mechanisms work by simply taking first-order representations from their respective functionally defined 'boxes', and then copying and embedding their contents into more complex higher-order representations. For instance, in the case of belief, the relevant monitoring mechanism copies the content that *p* from the belief box, embeds it in a representation schema of the form *I believe that* ____, and then puts the resulting higher-order representation that I believe that *p* back into the belief box (Nichols and Stich 2003, p. 161). Nichols and Stich believe that the same kind of monitoring mechanism can be operative for the case of desire, intention, imagination, and perhaps even perception, although the percept monitoring mechanism might be more complex (Nichols and Stich 2003, pp. 163–4).

However, as Goldman (2006, p. 239) rightfully remarks, Nichols and Stich leave completely unanswered *how* such monitoring mechanisms know which attitude type (belief, desire, etc.) they are dealing with. They do not suppose that there are any literal boxes to be found in the brain, from which the attitude type can then be immediately inferred by the monitoring mechanisms. But without such boxes, it is difficult to see how the monitoring mechanism could know in what representation schema a certain representational content should be embedded to end up with the correct higher-order belief. What is needed is an explanation of how monitoring mechanisms are able to distinguish between attitude types, only then can Nichols and Stich's theory about content embedding be successful.

[2] The phenomenon of known hallucination might be more difficult to deal with, although it's not obvious that hallucinatory subjects do not believe that they are perceiving in the relevant sense if they know that they are hallucinating. But note that I am here only presenting a suggestion to the metaphysical question of the phenomenal distinctness of perception; my epistemic position does not commit me to holding that there is a distinctive phenomenal property of experience that has to do with higher-order beliefs.

Now, given what I've said so far about the phenomenology of introspective knowledge, it should be clear that I take it to be implausible that monitoring mechanisms use phenomenal properties as input to distinguish between attitude types. According to Goldman (2006, pp. 247–253), this leaves functional, representational, or neural properties as possible inputs on the basis of which a monitoring mechanism could distinguish between attitude or mental state types. Functional properties would be difficult to detect, as they have to do with the effects a state *would have* given certain circumstances. Representational properties can be shared by many different attitude types, as they can have the same propositional contents: I can hope that *p*, fear that *p*, believe that *p*, etc. So Goldman concludes that neural properties are the best candidates on the basis of which a monitoring mechanism can identify the right attitude or mental state type.

This proposal can be correct even if mental states are not type-identical to physical states. As long as the different mental states are in some way connected to specific neural pathways, circuits, or activities, then a monitoring mechanism would be able to distinguish between them. But, in contrast to Goldman, we need not suppose that a monitoring mechanism only has access to *one* specific type of property. It seems that a monitoring mechanism would be far better at distinguishing between different mental state types if it used a combination of functional, representational, and neural properties. For instance, the level of detail of a representation (representational) in combination with information about its causal origins (functional) and the activated neural circuits (neural) could be sufficient to determine whether the representation was, say, a perception or imagination. Of course, one might hold that all these factors are, in the end, detectable by some sort of neural properties, thereby making the inputs to the monitoring mechanism of one kind again.

Notice, though, that we have so far focused mostly on the attitude types of perception and imagination, and for these attitudes a monitoring account of introspection certainly looks plausible. But for the more traditional propositional attitude types like belief and hope, it's not clear whether monitoring remains as good a theory. The idea that one could distinguish between a hope that *p* and a belief that *p* on the basis of neural properties appears rather speculative. Perhaps one would prefer the idea that we have only one *interpretative* system for the recognition of these propositional attitudes in ourselves and others, as proposed by Peter Carruthers (2011), rather than the idea that we have a separate inner sense by means of which we come to know about our own propositional attitudes. For our purposes though, the decision between these two competing views is not important for two reasons. First, as long as the cognitive process is reliable, the process reliabilist view will take its outputs to be justified no matter whether it is an interpretative or inner sense mechanism. Second, given that Carruthers also believes that his proposed interpretative mechanism is largely unconscious, an internalist view of introspection would remain problematic, and an externalist view seems best suited for the job.

So a process reliabilist view of introspective justification (for beliefs about perceptual states) is plausible from both the empirical and more theoretical point of view. It's plausible that we have a mechanism by means of which we can come to know that we are perceiving or imagining when we are doing so, and process reliabilism is able to provide the same account of justification for these higher-order introspective beliefs as it provides for the justification of first-order perceptual beliefs.

6.3.3 Alternative Views of Introspection

Of course there are also accounts of introspection that might appear entirely opposed to the kind of process reliabilist view advanced here. For instance, according to Horgan and Kriegel (2007), there are a certain type of beliefs that are simply *infallible*: one could not go wrong in forming those beliefs about one's current mental states. The type of beliefs Horgan and Kriegel are concerned with are so-called *SPPB phenomenal beliefs*: singular, present, phenomenal in mode of presentation, and bracketed beliefs about the phenomenal properties of one's experiences. Without going in too much detail, this type of belief can be expressed by a proposition of the form "This experience has *this* feature". Such a belief is a belief about one's current experience that employs the mode of presentations of both the experience and its relevant phenomenal feature to demonstratively refer to them. It is precisely because such beliefs employ these modes of presentations in this way that it seems one cannot go wrong in forming them.

One might think that such an infallibilist and constitutivist view of introspection is inherently opposed to the fallibilist process reliabilist view proposed here. In fact, though, the situation is not that bad. First, a process reliabilist view of introspection can in principle remain silent about the precise features in virtue of which introspection is reliable. All it needs to claim is that the introspective process delivers justified beliefs because it's reliable, the explanation of this reliability can then take on different forms and can even be compatible with full reliability for the case of SPPB beliefs. The process reliabilist view of introspection can even allow for different types of introspective processes that each produce beliefs with different degrees of justification.[3]

Second, note that Horgan and Kriegel restrict their infallibility claim to a specific type of introspective beliefs. Horgan (2012) makes clear that even beliefs like "I am now having a reddish experience" are not totally infallible because one might make a classification error with the result that the experience one now classifies as phenomenally reddish is not sufficiently similar to experiences that one has previously classified as phenomenally reddish. This means that the infallibility claim of the theory is severely restricted and certainly does not apply to the type of higher-order beliefs considered above, i.e., the belief that one *sees* that such-and-so is the case.

[3]Also see Schwitzgebel (2012) for more on the idea that introspection consists of a plurality of processes, although Schwitzgebel is also notoriously skeptical about the reliability of these processes.

Third, Horgan and Kriegel's account focuses on introspective beliefs about the phenomenal properties of one's experiences rather than introspective beliefs about the attitude-type of those experiences. Even if one were to have transparent access to the contents of one's experiences (which is already doubtful for reasons mentioned in the previous paragraph), then this would not provide one with access to the attitude type of the experience. But given that we usually do have beliefs about the attitude types of our own experiences (e.g., that it is perceptual), one surely is in need of an account that can also explain how these introspective beliefs can be justified.

So Horgan and Kriegel's account of introspection does not appear strongly opposed to the account of introspection proposed here. However, another constitutivist account of introspection, proposed by Declan Smithies (2012b), really is strongly opposed to a process reliabilist account. According to Smithies, introspective justification "is a distinctive kind of justification that one has to believe that one is in a certain mental state, which one has just by virtue of being in that mental state" (2012b, p. 261). This is opposed to a process reliabilist account of introspection, as such an account stresses, first, that being in a mental state is never sufficient for justification for a belief about that mental state (one also needs a reliable introspective mechanism to form that belief), and second, that one can have an introspectively justified belief that one is in a certain mental state without actually being in that mental state (if it is the output of such a reliable mechanism).

In opposition to process reliabilism, Smithies upholds the following *introspective accessibility thesis*:

> For some mental states M, necessarily, one is in M if and only if one has introspective justification to believe that one is in M and one thereby has an introspective way of knowing that one is in M.
>
> (Smithies 2012b, p. 263)

It's important to note that Smithies uses a propositional rather than doxastic notion of justification in a very extreme sense: according to Smithies one can have introspective justification to believe a proposition, or a way of knowing a proposition, even if one lacks the psychological capacities to form a belief in that proposition on the basis of these justifications or ways of knowing. Although Smithies thinks it is correct to hold that "the limitations of one's doxastic capacities do not impose corresponding limits on which propositions one has justification to believe" (2012b, p. 266), surely there are *some* limitations imposed by one's doxastic capacities. In what sense can a subject's mental state really be called accessible if that subject does not even have the capacity to form a belief about it? Similarly, in what sense can a way of knowing a proposition be called a way of knowing it if no one is even able to believe the proposition in question?

Smithies attempts to support his accessibility thesis by claiming that it "articulates an epistemic ideal, which abstracts away from contingent facts about the limitations of one's doxastic capacities" (2012b, p. 266). However, if one lacks a psychological process that enables one to form beliefs about one's own mental states, then one can be as ideally rational as one likes but one will never end up with introspectively justified beliefs about one's own mental states. Now, of course one could claim that we have the intuition that the ideal rational agent would always know his own mental states, and thereby try to derive the accessibility thesis about introspective justification, but we might as well claim that we have the intuition that the ideal rational agent knows all and only truths—thereby leading to an accessibility thesis which relates *truth* to having justification to believe. But not many would be willing to uphold such a strong condition on justification, i.e., that a proposition is true if and only if one has justification to believe that the proposition is true.

So there does not appear to be a very good motivation for Smithies' purely propositional version of introspective justification. However, Smithies also casts doubt on the intuitive adequacy of a reliabilist account of introspection:

> Consider the case of *hyper-blindsight*, which is just like super-blindsight [i.e., our case of Bill], except that the subject has a reliable mechanism that is disposed to generate noninferentially based higher-order thoughts about phenomenally unconscious visual states. Intuitively, the hyper-blindsighter does not have introspective justification to form beliefs about what is represented in her visual system any more than the super-blindsighter or the regular blindsighter does.
>
> (Smithies 2012b, p. 272)

Perhaps it's true that a hyper-blindsighter intuitively lacks introspective justification to believe that he is seeing such-and-so, given that he lacks any kind of visual experience. But this might have more to do with the connection between seeing and having visual experiences than the connection between reliability and introspective justification. In the usual sense of *seeing*, the hyper-blindsighter simply is not seeing that such-and-so is the case if he does not also undergo a certain visual experience. In that case, the hyper-blindsighter's higher-order mechanism probably will not count as reliable given that it will constantly output false beliefs about what the hyper-blindsighter is seeing.

However, there seems to be no reason to think that the hyper-blindsighter intuitively lacks introspective justification to believe, e.g., that his visual system is now processing such-and-such information. Although there is no phenomenal state that is at the basis of this introspective belief, it's not clear that this is required for all introspective justification. For instance, my belief that I remember that I went out for dinner yesterday is introspectively justified, but it's not clear that there is a certain phenomenal mental state on the basis of which I believe that I remember this. If someone were to challenge my belief that I actually remember this (instead maintaining that I am, say, merely imagining it), I would certainly be hard pressed to justify it on the basis of some sort of phenomenal mental state. Yet it seems strange to say that my belief that I remember that I went out for dinner yesterday is not *introspectively* justified.

So a process reliabilist account of introspection seems to come out on top, especially when one considers introspective beliefs about attitude types rather than attitude contents—which are the type of introspective beliefs that appear crucial for finding a fully satisfying response to the Clairvoyance Problem. In the following section this response will be worked out in more detail.

6.4 Higher-Order Beliefs and Defeat

In the previous section we've seen that process reliabilism can provide a plausible account of how higher-order beliefs about perceptual states are justified in a way that parallels its account of justification for first-order perceptual beliefs. Where the perceptual belief that *p* is non-evidentially justified because it arises from a specific kind of reliable perceptual process, the higher-order belief that I *see* that *p* is non-evidentially justified because it is the result of a specific kind of reliable introspective process. Once one has such a theory of justified higher-order beliefs at one's disposal, it can be used to strengthen the process reliabilist response to the problems of blindsight and clairvoyance in the way explained above. It can now be said that blindsighters and clairvoyants lack a type of evidential justification by means of factive reasons that is available to normal perceivers.

However, we've already encountered reason to think that this explanation does not go far enough. Unsophisticated cognizers always lack the proposed factive kind of evidential justification, yet they do not seem to have unjustified beliefs. What's more, blindsighters and clairvoyants simply appear to have beliefs that are unjustified tout court, instead of merely being evidentially unjustified. Fortunately, the process reliabilist has another way in which he can use our higher-order beliefs to respond to the Clairvoyance Problem. If higher-order beliefs are capable of *defeating* first-order beliefs, then the beliefs of blindsighters and clairvoyants can come out as ultima facie unjustified even if they were produced by a reliable cognitive process.

On first glance there appear to be two ways in which a reliabilist might use higher-order beliefs to get at this result. The first option is to posit a higher-order belief on the part of the blindsighter or clairvoyant with the content that one was not perceiving that such-and-so is the case in the blind field, or that one was not perceiving that the President is in New York City. Such a higher-order belief could then cast doubt on the truth and justification of the first-order belief. If one feels that this belief is not sufficient to constitute a defeater, then one might also posit, as a variation on this first option, the higher-order belief that one does not know the source of the first-order belief, or the higher-order belief that the first-order belief was unreliably formed. Whichever the precise details, the idea is that the posited higher-order beliefs should conflict with what the relevant subjects themselves believe to be required for having justified beliefs.

The second option is to just use the *lack* of a higher-order belief about, e.g., the source of one's first-order belief as a defeater for the justification of that first-order belief. Although this second option is clearly less cognitively demanding, thus making sure that defeat is not too hard to come by, it also has the drawback of making defeat too easy to come by. Just think once more of cognitively unsophisticated agents that are unable to form any higher-order beliefs. For them, all perceptual justification would be defeated according to the second option. This makes the first option the preferable one. Having the relevant defeating higher-order belief appears demanding enough to ensure that perceptual justification is not always defeated for unsophisticated agents, but still appears easy enough to make sure that the justification is defeated for the more sophisticated agents in the relevant counter-examples to reliabilism.

On this proposal, higher-order beliefs will perform double duty for the reliabilist. First, they can provide additional evidential justification for believing that *p* by supplying factive reasons for those beliefs. Second, they can defeat the justification of first-order beliefs by casting doubt on their trustworthiness. However, one might have some doubts about this second role of higher-order beliefs, especially when one considers what this would mean for the introspective mechanisms that give rise to them. It seems likely that our higher-order mechanisms are not just meant only to monitor our first-order mechanisms and output beliefs about them. Instead, they surely also have the function of *controlling* first-order mechanisms. But if this is true, then it seems that higher-order mechanisms should prevent first-order beliefs from arising altogether instead of just outputting higher-order beliefs about the sources of these first-order beliefs. In other words, it seems likely that the justification of first-order beliefs is not just defeated by higher-order mechanisms, the first-order beliefs will probably never even reach the status of belief in the first place because of the higher-order mechanisms.

So process reliabilism will have to provide a somewhat more developed story about what goes on in the relevant cases of blindsight and clairvoyance. This will also provide an opportunity to give an account of defeat that does not rely too much on the notion of *reasons*, given that process reliabilism was meant to be a fully reductive theory of epistemic justification. Appealing to defeating higher-order beliefs might already have been going too much towards a non-reductive account of justification in the eyes of the radical reliabilist. We will now turn to discuss several ways in which a reliabilist might account for the type of higher-order defeat at issue in the problematic clairvoyance-style cases raised against it.

6.4.1 Defeat by Alternative Reliable Processes

One way in which we can characterize defeat from a reliabilist perspective goes back to Goldman's (1979) idea that defeat is related to alternative reliable processes:

ARP *S*'s belief in *p* at *t* is defeated if and only if there is an alternative reliable or conditionally reliable process available to *S* which, had it been used by *S* in addition to the process actually used, would have resulted in *S*'s not believing *p* at *t*.

Perhaps by using ARP, we will be able to accommodate the clairvoyance-style cases in terms of defeat. Now, recall that Goldman (2011a) also appealed to defeat to explain clairvoyance-style cases like that of Truetemp. According to Goldman, Truetemp should have been able to recognize that his beliefs popped up out of the blue, thereby arriving at a defeater for his belief. However, the problem with this account was that it was difficult to see how this defeat could be cashed out in terms of ARP. Indeed, if Truetemp had used introspection in combination with reflection, he might well have decided that he should not hold on to his belief, but such an alternative extended type of process should not be able to defeat justified first-order beliefs if it was not actually used. If this type of alternative process defeats the justification of beliefs, then it seems that too many of our beliefs would be defeated. By extensive reflection I might derive contradictions from a certain set of my beliefs, but if this derivation is not obvious then the justification for these beliefs should not count as being defeated.

With our new process reliabilist account of introspection, Goldman's suggestion could be made to work though. Suppose that we do have reliable introspective mechanisms that output not only beliefs about mental contents and attitude types, but also check the trustworthiness and adequacy of incoming information and the way it's being processed. Such *monitoring mechanisms* constitute alternative reliable processes that, had they been used in addition to the respective clairvoyant processes, would have resulted in the clairvoyant subjects' no longer believing as they do. After all, the information that the clairvoyants receive is not strongly corroborated by any other available information, and it also lacks any recognizable trustworthy source. This should be enough to lead the monitoring mechanisms to prevent this information from rising to the status of belief. Furthermore, they would do so without thereby rendering all kinds of first-order beliefs defeated: first-order beliefs stemming from known and trustworthy sources would not be rejected by the supposed monitoring mechanisms.

The appeal to such monitoring mechanisms also helps to rebut an argument against Goldman's original idea of accommodating clairvoyance in terms of defeat. According to Jack Lyons, Goldman's suggestion is doomed to fail because one can stipulate that Norman's reflective capacities are so bad that he has no available *reliable* mechanism to defeat the justification of his clairvoyant beliefs (Lyons 2009, p. 124). But if we appeal to monitoring mechanisms that supposedly constitute an important part of human cognition, then it will no longer be easy to just stipulate that these mechanisms are unreliable. The consequences of such a stipulation would no longer be clear, especially if, as suggested in Sect. 6.3.1, the output of such monitoring mechanisms is relevant to the phenomenology of our perceptual experiences. This will make it much more difficult to give an intuitive evaluation of the posited subject and his beliefs, at least when the details are spelled out completely.

However, even if this story about monitoring mechanisms is correct, then it will still be difficult to successfully integrate with ARP and process reliabilism. First, with regard to the reliability of the monitoring mechanisms, given that the mechanisms are supposed to *reject* beliefs, reliability cannot be accounted for in terms of a tendency to produce true rather than false beliefs. Perhaps this can be solved by focusing on the fact that monitoring mechanisms *prevent* the formation of false beliefs, while not preventing the formation of true beliefs. Monitoring mechanisms can then be said to be reliable if they prevent sufficiently more false beliefs from occurring than they prevent true beliefs from occurring. But even if such a suggestion is on the right track, there is still a second and third problem left.

The second problem with combining ARP and monitoring mechanisms is that it's not really clear whether the monitoring mechanisms should be seen as *additional* mechanisms or as *parts of* our perceptual mechanisms. Either option will lead to difficulties in responding to the clairvoyance-style cases against process reliabilism. If we suppose that the monitoring mechanisms are additional to our perceptual mechanisms, then, given the set-up of Truetemp's case, we can suppose that the monitoring mechanism is being overridden by the tempucomp. This means that Truetemp's belief is the output of a reliable mechanism with no *available* alternative reliable process to defeat its justification. In other words, Truetemp still comes out as having a justified belief.[4] If, on the other hand, we suppose that the monitoring mechanisms are a *part* of our perceptual faculties, then Bill, Norman, and Truetemp should all be seen as using entirely different processes to get at their respective beliefs. But in that case, there would again be no alternative available mechanism that would lead to the rejection of their beliefs.

Now, there might be a way around this second problem by using a rather different notion of availability according to which a cognitive process is available if a subject is generally able to use it (even if the subject is not able to use it in this particular instance). But there is still a more fundamental problem with ARP itself left: it's not clear that it can be properly motivated from within a process reliabilist framework. Recall that process reliabilism aims to provide a reductive account of justification in terms of reliability. Supposing that justification really is reducible to, or supervenient on, the reliability of a belief-forming process, then why does it matter that beliefs might have turned out differently if another reliable process was used? Consider the following example: I have two textbooks available that are both equally and highly reliable, and I happen to pick one of them to find out more about a certain topic. If the textbook is reliable about its subject matter, and justification is reducible to reliability, then it shouldn't matter what the other, equally reliable and available textbook, would have told me. So it seems that one cannot uphold ARP as a defeat clause if one also wants to maintain, with process reliabilism, that justification really is *reducible* to the reliability of belief-forming processes.

One response on behalf of process reliabilism is that its claim is not that justification is reducible to just the reliability of the actual belief-forming processes, but also to the reliability of processes "...that could and should be employed" (Goldman 1979, p. 20). But this does not yet answer the question as to the motivation of this addition. One such motivation might be that an epistemic agent has not done her epistemic duty if she neglects to use a very reliable method at her disposal, but this just raises a question as to whether reliabilism can then still count as having provided a fully *reductive* account of justification. Epistemic duties should be explained in terms of reliability, rather than using them as primitives to explain which forms of reliable processes (actual and potential) should be used to account

[4] Note that Bob Beddor (2015, pp. 155–6) also uses these problems with the availability of alternative cognitive mechanisms to argue against the idea that having such available mechanisms is necessary for defeat.

for justified beliefs. What's more, this account of epistemic duty also seems to become overly demanding if an agent is always expected to use two methods if they are equally reliable, instead of just using one of them.

It might be helpful to compare this complaint to one raised by Thomas Grundmann (2009). According to Grundmann, ARP is ad hoc as it "fails to explain why internally rational counterevidence removes justification" (Grundmann 2009, p. 70). Grundmann seems particularly worried about the fact that this counterevidence can be false and unreliably produced, thereby defeating beliefs that are themselves true and reliably produced. Goldman responds to this worry in part by saying that he doesn't think "it's the responsibility of a principle to explain *why* satisfaction of the defeater-sensitivity condition should be correlated with improvement in overall truth-ratio, as long as it *is* so correlated" (Goldman 2009, p. 248). Now Goldman is probably right in thinking that satisfaction of his defeater-sensitivity condition is actually correlated with an improvement in overall truth-ratio. This could provide a *general* motivation for the defeat-clause even though forming beliefs in accord with this clause could lead to a lower truth-ratio in *particular* instances (e.g., cases in which the counterevidence is itself unreliably produced). But this still would not answer the worry I have raised. This worry has to do with the fact that process reliabilism appears committed to holding that a certain amount of process reliability is sufficient for justification, *irrespective* of whether other processes would increase this reliability even more.

These difficulties provide us with sufficient reason to look at a different account of defeat. For instance, a defeat clause in proper functionalist terms goes a long way in circumventing the above problems: such an account can supply a specific proper function for defeat mechanisms, has no problems with accounting for the availability of these mechanisms, and can be properly motivated from within the proper functionalist framework. We will now turn to explore this suggestion in more detail.

6.4.2 Proper Functionalist Defeat

The idea of this proper functionalist account of defeat makes use of Plantinga's suggestion that we have a "defeater system" (Plantinga 1993, pp. 40–42) which is aimed at forming true beliefs when functioning properly:

> The basic idea here is that the design plan is such that (for example) when you are appeared to a certain way you will form a certain belief: when you are appeared to redly, you will (ceteris paribus) form the belief that you see something red. But the design plan also specifies circumstances under which, even though you are appeared to redly, you won't or don't form that belief. These circumstances would include, for example, your learning that the thing in question, despite appearances, is not red (rebutting defeater), or your coming to believe that the thing would have looked that way even if it were not red (undercutting defeater).
>
> (Plantinga 1993, p. 41)

As we saw before, one of the problems with Plantinga's account is that it makes use of the notion of a "design plan" to explain the proper function of cognitive systems. Not only does that create difficulties with the case of Truetemp, where the *designed* Tempucomp reliably outputs temperature beliefs which are nevertheless unjustified, it also has problems with accounting for justification by newly acquired cognitive processes that lack any design plan. Instead, I prefer the framework provided by Graham (2014) in which proper function has to do with how a system performed when it produced the beneficial effects that led to that system's continuing existence.

Now, as we saw in the previous chapter, the specific proper function that Graham takes to be necessary for producing justified beliefs is the proper function of reliably producing true beliefs. It seems that a similar reliability-related function should be connected to defeat, although it should allow for the fact that defeater systems are in the business of *preventing* false beliefs rather than producing true beliefs. A good suggestion here is that defeater systems are those systems that have the proper function of reliably preventing the formation or maintenance of false beliefs (without thereby preventing the formation or maintenance of true beliefs).[5]

In contrast to process reliabilism, proper functionalism can also give a good account of the relevance of such defeater systems. Given that the output of our cognitive faculties are influenced by what the defeater systems have assessed as trustworthy, our cognitive faculties are no longer selected purely on the basis of their own merit, but in combination with the merit of the defeater systems. Thus, the consequence etiologies for our cognitive faculties by now seem to *include* our defeater systems in an important way. This means that these defeater systems cannot be ignored when considering the full justification of our beliefs.

A proper functionalist account of justification can thus make a motivated appeal to defeater systems that have their own proper function relevant to justification. But a question that still remains is how this defeat should be spelled out exactly. According to Michael Bergmann, no beliefs are justified if a subject's defeater systems are not functioning properly (Bergmann 2006, p. 170). This, however, seems too strong to work. Recall that on the proper functionalist proposal under consideration, one's perceptual systems have the proper function of reliably producing true beliefs. So even if one's defeater systems are not functioning properly, then one's perceptual systems will still reliably produce true beliefs if they are functioning properly. On Bergmann's proposal, if one's defeater systems are somehow malfunctioning, then such reliably produced true beliefs will not be justified—even if a properly functioning defeater system would not have rejected them. And this appears to be an unwelcome result: perceptual faculties are now deprived of their justificatory power even if the beliefs they produce are perfectly fine. That is why I propose a slightly weaker Proper Functionalist Defeat (PFD) clause, in line with suggestions by both Grundmann (2009) and Bedke (2010):

[5]This suggestion is made in slightly different terms of developmentally moderating processes by Bedke (2010, p. 7).

PFD *S*'s belief in *p* at *t* is justified only if *S* does not have a defeater system *D* such that, had *D* been working properly, it would have resulted in *S*'s not believing *p* at *t*.[6]

Note that this proper functionalist defeat clause no longer encounters the availability problem that ARP had in the classic process reliabilist framework. A belief will be defeated if there is a defeater system present that would have resulted in the rejection of this belief if it had been working properly. So even if the defeater system is not actually available due to circumstances that prevent it from doing its work, what matters is what the defeater system would have done if it had been functioning properly.

PFD also allows for cases in which beliefs are both true and reliably produced, and yet still defeated because they would not have been formed if the defeater-system had been working properly. But according to a proper functionalist account, this is exactly how it should be, given that defeater systems have become a part of our total cognitive system. Even though defeater systems might get it wrong in a particular case, their contribution to our overall cognitive structure is still important enough to validate their decisions even in those rare cases where they get it wrong.

Indeed, this is exactly what goes on in Clairvoyance-style cases.[7] Notice that the earlier proposed monitoring mechanisms count as defeater systems as they have been explained above: they check incoming information for trustworthiness, and have presumably been selected because this reliably prevents false beliefs from occurring. They thus have what it takes to constitute a defeater system in the proper functionalist sense.

Given that monitoring mechanisms are a specific kind of defeater system, we will need to look at what these monitoring mechanisms would have done if they had been functioning properly to see whether the beliefs of blindsighters and clairvoyants are defeated. Fortunately, it certainly appears plausible that these monitoring mechanisms would have rejected their respective beliefs had they been functioning properly. For instance, Bill knows that he is not experiencing anything in a certain part of his visual field, which should undermine the information that such-and-so is going on in this blind field. This would also explain why regular blindsighters in fact do not form any beliefs about what goes on in their blind field, but have to be forced to guess. So PFD can explain why Bill's beliefs are unjustified in line with the Blindsight Intuition.

---

[6]Note that this type of defeat clause will also circumvent the counterexample to Bergmann's theory provided in Johnson (2011), as it only deems the proper functioning of defeater systems relevant if they would have prevented a belief from occurring had they been functioning properly.

[7]Colin Ruloff (2000) also argues that an appeal to defeat can protect Plantinga's proper functionalism from some of BonJour's counterexamples. However, to account for the precise way in which defeat occurs Ruloff stresses too much (as Plantinga sometimes does (Plantinga 1996, p. 377)) that we already believe that there can be no such thing as clairvoyance. My account is able to explain how defeat arises without supposing that we have any beliefs of this sort.

Now one might object to this defeat-account of the Blindsight Intuition on the grounds that it does not do full justice to it. According to this objection, Bill intuitively lacks perceptual justification because he lacks experiential evidence. Any explanation that does not appeal to the lack of this experiential evidence thus fails to account for the Blindsight Intuition in a satisfying way. However, this seems to be an exaggeration of what our intuitions about justification amount to. For instance, if we add the fact that Bill knows he is a fully reliable blindsighter to our original blindsight case, then our intuition that he forms unjustified beliefs disappears even though his experiential situation has remained unchanged. So we are better off by simply taking the Blindsight Intuition as one that claims that Bill has unjustified beliefs, thereby allowing for different kinds of theoretical explanations for this intuition.

For the hypothetical cases of Norman and Truetemp, the situation is somewhat different. In contrast to Bill, both Norman and Truetemp lack any previous acquaintance with their respective senses, and so do not have any background information about them. However, exactly for this reason it seems that the information they provide will not be simply accepted by their monitoring mechanisms. The information that is outputted by the new senses is not corroborated by any of the other senses, the source of this information is unknown to the monitoring mechanisms, and there is also no established way of how this information should be integrated and calibrated with the more usual kind of sensory information. This should be sufficient to prevent this information from rising to the status of belief, although further use could certainly end in the new senses being fully integrated with both the monitoring mechanisms and the other senses.[8]

In effect, I am assuming that a belief-forming mechanism will have to be used several times to be successfully integrated with the monitoring mechanism as a trustworthy source. However, I am not taking this etiological condition on belief-forming mechanisms as a necessary condition on justification, but just as a contingent condition on justification that is the result of our precise cognitive make-up as human epistemic agents. This derives some support from the findings presented by Beebe (2004), which show that in actual cases where subjects receive new perceptual faculties (such as cochlear implants, tactile vision, etc.), these subjects normally do not immediately start to form beliefs on the basis of that faculty. Of course, in such actual cases there is also the matter of *calibrating* the new senses with the old ones, something that is completely ignored (or assumed to be unnecessary) in all of the hypothetical cases we're discussing. In any case, if we assume that monitoring mechanisms can recognize new perceptual faculties as trustworthy before they have ever been used, then some of the intuitive force of

[8]Note that the way in which I propose that defeat occurs for Norman is importantly different from the way in which defeat occurs in BonJour's other clairvoyant cases (BonJour 1985, pp. 38–40). The other clairvoyants all believe themselves to have a faculty of clairvoyance, even though they all have some evidence that no such faculty is possible or reliable. Norman, in contrast, does not have any evidence of the sort, and also does not believe himself to have a clairvoyant faculty. Yet still his clairvoyant beliefs are defeated, I propose, because of higher-order mechanisms that should have kept him from trusting the 'strange' incoming information.

the examples might change when they are properly re-described. For instance, the beliefs of the respective subjects might no longer seem to pop up out of the blue if they are accommodated by a higher-order belief that they stem from a trustworthy source.

With regard to the case of Truetemp, there is another worry. The case of Truetemp stipulates that the tempucomp leads to Truetemp's unreflective *acceptance* of temperature beliefs, so a monitoring mechanism might not even come into play. But even if this is stipulated, then the current framework need not imply that these beliefs are now justified. The question we need to ask is whether the tempucomp's *overriding* Truetemp's monitoring mechanisms somehow impacts the proper functioning of these mechanisms. And there is reason to think that this proper functioning is indeed impacted negatively. In such a case, the monitoring mechanisms are definitely not functioning in the way they were when they got selected, as they are not monitoring a specific type of incoming sensory information. And if this is true, then what is relevant is what they would have done had they been functioning properly, namely: preventing this strange sensory information from reaching the status of full-blown beliefs.

Finally, note that this answer to the clairvoyance-style counterexamples to reliabilism does not have the unwanted consequence that unsophisticated animals or small children lack justification for their perceptual beliefs—although the explanation of this fact is slightly different for each of these groups of agents. Given that monitoring mechanisms are not necessary for perceptual justification (as they are only necessary for a specific kind of defeat), even unsophisticated animals which lack such mechanisms can still have justified beliefs. Small children, in contrast, presumably have monitoring mechanisms that have not yet developed fully. However, this is perfectly *proper functioning* for these monitoring mechanisms. This means that some beliefs that are defeated for adults will not be defeated for these small children. Indeed, if Norbert's clairvoyant faculty would already have been active during his early childhood, then these early clairvoyant beliefs might well be justified.[9] However, this is not a weakness but rather a strength of the theory, as this accords with our intuitions about such matters: given that adults have more sophisticated cognitive capacities than small children, adults have to satisfy higher demands to have ultima facie justified beliefs.

6.4.3 PFD and Process Reliabilism

The question that is left is whether PFD commits us to a full proper functionalist framework, or whether it can be combined with the more classic process reliabilism.

---

[9]Note that his later beliefs will then also become justified, as his monitoring mechanisms would then presumably accept his later beliefs as stemming from a trustworthy source due to their good track record.

This question might be thought especially pressing as I have already argued against some aspects of proper functionalism in the previous chapter. Note, though, that the aspects I have argued against are, first, the *necessity* of proper function for justification, and, second, the use of proper function to account for the absence of *prima facie* justification in the Clairvoyance Problem. The current proposal to solve the Clairvoyance Problem in terms of proper functionalist defeat seems compatible with both of these two critiques. It acknowledges that the Clairvoyance Problem is about ultima facie rather than prima facie justification, and its motivation relies on the thought that having a specific sort of proper function is *sufficient* for having an impact on prima facie justification rather than this function's being necessary for prima facie justification. It is true that, combined with a variety of process reliabilism (which would also make a specific kind of reliable process sufficient but not necessary for justification), this would make justification into a property that could be achieved by different types of processes, but it is not immediately clear that this is a devastating rather than inelegant outcome.

Briefly put, the rationale behind such a view could be as follows. Our epistemic goal is to achieve true belief, and processes that are connected to reliably achieving this goal are evaluated positively in the sense that the beliefs they produce are prima facie justified. However, there are different *ways* in which processes could be connected to the goal of reliably achieving true belief. For instance, processes could either actually reliably produce true beliefs (the focus of process reliabilism), but they can also work in the way they did when they actually reliably produced true beliefs in the past (the focus of proper functionalism). Although the latter way of performing does not guarantee that the process is currently reliably producing true beliefs, it seems like an epistemically proper way to perform given that it has already been successful in the past. In this way the property of being a justified belief might be realizable by different types of processes.

However, even for those who are reluctant to accept such a disjunctive view of justification there appears to be a possibility to combine PFD with process reliabilism. Given that higher-order defeat can only occur if first-order beliefs are already prima facie justified, the situations we are concerned with, seen from the process reliabilist's perspective, are ones in which the first-order processes are already reliably producing true beliefs. All that's needed now is a process reliabilist motivation for looking at a counterfactual condition in terms of properly functioning higher-order monitoring mechanisms. This motivation can be given by realizing that, if the counterfactual clause is fulfilled (that is, if the first-order belief would have been rejected had the monitoring mechanism been working properly), then we have a good indication that the first-order perceptual beliefs are currently being *sustained* by an *unreliable* process: namely, the process by means of which the higher-order monitoring system is letting strangely caused beliefs go unchecked. In contrast to Goldman's defeat clause, which focuses on the availability of an alternative reliable process that would have led to a different belief, on the current suggestion we have to account for defeat by focusing on there being an unreliable process that is actually used to *sustain* the current belief. If PFD is fulfilled, then that is a good indication that such an unreliable process is present.

Note that this proposal does not fall prey to the earlier objection of ad-hoc-ness. It still fully adheres to the idea that justification supervenes on the reliability of cognitive processes; it just adds the thought that which processes *sustain* a belief also matters for that belief's ultima facie justification. This idea is true to the spirit and the letter of the reductive framework provided by classic process reliabilism.

However, one might still doubt whether the current approach to defeat can really be upheld. The approach is one that Jonathan Kvanvig (2007) has labeled as a 'backdoor theory' of defeat, as it characterizes defeat in terms of beliefs that are kicked out of the backdoor of the cognitive system instead of characterizing defeat in terms of the propositional contents that arrive at the front door. In this critique, Kvanvig specifically targets Plantinga's (2000) account of defeat*ers*, beliefs or experiences that defeat the justification of other beliefs. According to Kvanvig, this account will have difficulties accounting for, first, defeater-defeaters, and, second, the *several* rational possibilities one has in responding to defeaters, as a backdoor approach only looks at belief rejection. Instead, the propositional approach to defeat is supposed to fare better, as this can provide simple and illuminating definitions of defeaters and defeater-defeaters in terms of what is (no longer) supported by the evidence:

> Where *dd* is a defeater of *d*, the conjunction of any evidence *e* conjoined to *d* does not justify *p*, but the conjunction of *e* plus *d* plus *dd* yields at least as much justification for *p* as provided by *e* itself. Moreover, the Quine/Duhem problem ceases to worry as well, for even if *d* is a defeater of the $p|e$ relation (where *e* is the evidence for *p*), it need not be a defeater of the $p\&r|e$ relation. That allows rational adjustments to a system of beliefs in response to learning *d* that don't require abandoning *p*.
>
> (Kvanvig 2007, pp. 119–20)

However, process reliabilists and proper functionalists have good motivations not to accept this picture of defeat. Propositional relations by themselves won't provide a satisfying theory of defeat, as what counts as a defeater (or evidence for that matter) for a belief that *p* for one agent is not the same as what counts as a defeater (or evidence) for a belief that *p* for another. One agent might see many consequences of his beliefs more easily than another; this means that this former agent will encounter inferential defeat more easily than the latter. That is why ARP focuses on *available* reliable processes, and why PFD also focuses on possibly individually *developed* defeater systems: to account for the possible differences in defeat between different epistemic agents. So process reliabilists and proper functionalists have good reason to focus primarily on the cognitive processes by means of which beliefs are defeated, rather than the propositional contents of defeating and defeated beliefs. This will make the notions of 'defeater' and 'defeater-defeater' only of secondary interest as they tend to abstract away from the specific cognitive processes of specific agents. What's more, if a subject *S*'s defeater system would not have led to the rejection of a certain belief *B* (but would have opted for a different rational response), then *B* simply is not defeated for *S* according to the theory, even if the defeater system of another subject *S*′ would have rejected *B*.

This again underlines the idea that defeat can be different for each subject, and that it is an idealization to speak of defeaters independently of specific subjects.

6.4.4 Comparing Similar Theories

The proposed answer to the Clairvoyance Problem might in some ways appear similar to the answer proposed by Ernest Sosa (2007) and the answer proposed by Daniel Breyer and John Greco (2008). I'll briefly take the time to discuss the differences between these answers and the one provided here.

According to Sosa, the problem with clairvoyance comes down to a difference between "animal" and "reflective" knowledge (Sosa 2007, p. 24). Where animal knowledge merely requires a subject to truly believe that p because of a certain competence or ability to form true beliefs, i.e., an *apt* belief on the part of the subject, reflective knowledge also requires the subject to aptly believe that one's first-order belief is apt.[10] Applied to the cases of clairvoyance, on Sosa's view we get the result that the subjects in these cases lack reflective knowledge even though they possess animal knowledge of the relevant propositions. Norman aptly believes that the President is in New York City, but he does not aptly believe that he aptly believes that the President is in New York City.

Now one might think that this account of clairvoyance is similar to the defeat-account I have proposed above. After all, according to the defeat-account, unsophisticated epistemic agents, "animals", can have justified beliefs without having any higher-order beliefs as long as their first-order beliefs are the result of a certain sort of reliable process. In addition, on the defeat-account, sophisticated epistemic agents, "reflective agents", normally have justified higher-order beliefs about their own first-order beliefs that tell them whether they stem from a reliable source. So why not accept that there are two different kinds of justification, one animal, which does not require higher-order beliefs, and one reflective, which does require such higher-order beliefs?

The answer is that the defeat-account can simply accept that there is only one kind of justification, but that its clause for (higher-order) defeat comes into play only for the more sophisticated epistemic agents. According to the defeat-account of clairvoyance, unsophisticated agents are actually better off than sophisticated agents in the relevant cases of clairvoyance: where unsophisticated agents lack the monitoring mechanisms required for higher-order defeat, and thereby acquire justification and knowledge when they are in a clairvoyance-style case, reflective agents do have

[10] I limit myself here to a very brief discussion of the distinction between animal and reflective knowledge, although of course much more could be said about Sosa's theory in general. Note also that Sosa has developed his views further in ways that are not crucial for the current discussion (Sosa 2011, 2015).

monitoring mechanisms at their disposal, and are thereby prevented from acquiring knowledge and justification in clairvoyance-style cases. Of course this does not mean that unsophisticated agents are all-things-considered better off than sophisticated ones. Presumably our reflective capacities bring with them an overall higher degree of reliability and other epistemic bonuses. In contrast, on Sosa's account both types of agents, unsophisticated as well as sophisticated, have animal knowledge but not reflective knowledge in the clairvoyance-style cases. So the proposed account of clairvoyance is importantly different from the one proposed by Sosa.

Breyer and Greco (2008) have presented an account of clairvoyance that is also in some respects similar to the one proposed here. Their proposal builds on Greco's earlier *agent reliabilism* (Greco 2000), according to which reliabilism should focus more on the agent who uses certain cognitive processes rather than focusing too much on those processes themselves. This leads them to the following suggestion about clairvoyance-style cases:

> Brain lesions and clairvoyance (of the proposed sort) do not count as part of agent character, and precisely because they are not well enough *integrated* with other of S's cognitive dispositions. Lacking such integration, clairvoyance looks like a sub-personal mechanism rather than an agent ability or power [. . .] Accordingly, the reliability of these mechanisms does not amount to agent reliability, and so their outputs do not get a positive evaluation on that score.
>
> (Breyer and Greco 2008, p. 175, my italics)

The important notion in this response to clairvoyance is that of *cognitive integration*. Breyer and Greco develop this notion in several ways, the clearest of which appeals to a subject's reflective endorsement of his beliefs as stemming from a certain reliable cognitive disposition, or to a subject's taking responsibility for his beliefs as his own (Breyer and Greco 2008, pp. 179–181). Now, although it certainly seems true that all of the clairvoyant subjects do not either reflectively endorse their beliefs as stemming from reliable cognitive dispositions, or take responsibility for their beliefs as their own, these higher-order requirements face similar worries as those originally proposed by BonJour and Lehrer themselves. Unsophisticated cognizers are *always* unable to reflectively endorse or take responsibility for their beliefs, yet they do seem capable of having perceptual justification. The notion of cognitive integration will thus have to be spelled out differently if it is to provide a satisfying answer to the Clairvoyance Problem.

This leads me to another of Breyer and Greco's suggestions, which appeals to *structuralism* to explain the notion of cognitive integration. According to structuralism, for a belief to be integrated it needs to be the result of "the cooperative causal interaction of relevant cognitive dispositions" (Breyer and Greco 2008, p. 183) and cohere sufficiently with other of the agent's beliefs. The problem, though, is that it is not really clear what these requirements amount to. Some of our beliefs are relatively independent from others (think of, e.g., memorial beliefs such as that Hector was slain by Achilles), and clairvoyant beliefs might just fall into this category. The most plausible way of spelling out the structuralist position will probably have to say something about how first-order beliefs are integrated with higher-order mechanisms that calibrate them with other information in the cognitive system. But, again, if this is meant to be a necessary condition on justification, then unsophisticated cognizers will turn out to have unjustified beliefs.

Appealing to defeat thus appears to be the best way to go.[11] This allows us to have our cake and eat it too. We can explain what goes on in cases of clairvoyance by pointing to the higher-order mechanisms of the relevant subjects, while it does not fall prey to the problem of hyper-intellectualization. This means that we are able to explain the continuity between unsophisticated animals and human cognizers with respect to first-order justification (i.e., reliable processes of a certain sort), while also preserving the idea that there is something special about sophisticated cognizers that are capable of higher-order thought (i.e., their first-order justification can be defeated by higher-order processes).

6.5 Explaining the New Evil Demon Intuition

The proposed answer to the Clairvoyance Problem made use of the idea that first-order perceptual beliefs could be defeated by higher-order monitoring mechanisms. A similar explanation is obviously not available for the New Evil Demon Intuition, as subjects misled by an evil demon intuitively do have justified beliefs. However, one way in which the underlying theory of perceptual justification could be cashed out, i.e., in terms of a hybrid of proper functionalism and process reliabilism, also provides a good way of accommodating all worries about the New Evil Demon.

In the previous chapter we saw already that proper functionalism had a promising response to the standard New Evil Demon scenario. Given that Ned's perceptual system is functioning properly with the aim of reliably producing true beliefs, Ned's perceptual beliefs turn out to be justified. After we've now ascertained the possibility of a process reliabilist way of grounding higher-order factive reasons, we can add that proper functionalism even has another benefit in its response to this standard version of the New Evil Demon scenario. Not only are Ned's perceptual beliefs non-evidentially justified because they are the output of a properly functioning cognitive mechanism, they can also be evidentially justified by higher-order (but, in this case, non-factive) reasons. Ned's introspective mechanisms are presumably still functioning properly, that is, they are presumably still functioning in the way they were when they got selected in Ned's ancestors.[12] In that case, Ned's higher-

[11]Note that Greco (2003) also mentions sensitivity to defeating evidence as yet another aspect of cognitive integration. The problem is that he does not explain why Norman, Truetemp, or Norbert would be insensitive to any defeating evidence, a complaint also made by Sven Bernecker (2008).

[12]Of course this depends on how the demon is intervening precisely. But I think that the scenario implicitly assumes that the demon is not *changing* any of the subject's psychological processes, it's just stimulating his sensory faculties in a way that is normally done by external stimuli from the world.

order beliefs that he is now perceiving such-and-so come out justified, and they are capable of providing further evidential justification for his first-order beliefs.[13]

However, in the Reversed Evil Demon scenario both these advantages of proper functionalism become disadvantages. Think back to Red, the subject in this reversed scenario. Red has been deceived by an evil demon since the beginning of his life, and, because of this etiology, both his perceptual and introspective mechanisms will no longer have the proper function required to produce justified beliefs, i.e., that of reliably producing true beliefs. When Red is unknowingly released from the evil demon's grasp, he will simply continue to form his first-order and higher-order beliefs the way he did when he was still under the demon's influence. Given that these beliefs will now be reliably formed in exactly the same way that we form our beliefs, it seems that Red should have justified first-order beliefs and justified higher-order beliefs. Red can now simply see that such-and-so is the case, and he now also appears capable of knowing that he sees that such-and-so is the case. But, unfortunately, this intuition cannot be captured by proper functionalism, as Red's cognitive mechanisms no longer have the right kind of proper function due to his having been radically deceived all his life. This means that Red will lack non-evidentially justified perceptual beliefs, and will also lack the required justified higher-order beliefs by means of which he could further evidentially justify his perceptual beliefs. So although proper functionalism appears to provide a promising solution to the New Evil Demon scenario, its response remains problematic for the Reversed Evil Demon scenario.

If we allow a hybrid of proper functionalism and process reliabilism though, even the Reversed Evil Demon scenario comes out correctly. According to this approach, both proper functioning *and* a specific type of process reliability can be sufficient for prima facie justified beliefs. True, Red lacks a perceptual system that has the proper function of reliably producing true beliefs, but he does have a perceptual system that currently *is* reliably producing true beliefs. Given that Red also lacks a monitoring mechanism that has the proper function of reliably preventing false beliefs, these reliably produced perceptual beliefs will also remain undefeated. So Red turns out, in line with our intuitions, to have undefeated justified perceptual beliefs.

As I have mentioned before, a hybrid view of justification will no doubt be relatively controversial to accept though. That is why I also want to set out a different type of response to the New Evil Demon Intuition, one that is in line with our focus on the possibility of non-evidentially justified perceptual beliefs. Instead of accommodating the truth of the New Evil Demon Intuition, this response aims to explain why we would mistakenly have such an intuition in response to the scenario.

The proposed explanation starts from the fact that it is commonly assumed that we should account for the justification of perceptual beliefs in terms of *evidential* justification, where the relevant evidence is constituted by perceptual experience.

[13]Note that Graham (2012, p. 455) also mentions that demon victims can provide justifications for their beliefs, but he does not explain how these justifications could depend on further *justified* beliefs.

This experientialist assumption seems to be accepted by philosophers of different stripes, internalists as well as externalists, so there is good reason to think that it might play a role in our intuitions about the New Evil Demon scenario.[14] Now, if it is true that we base our perceptual beliefs on experiential evidence, then, given that radically deceived subjects like Ned have the same perceptual experiences as us, they seem to have the same evidence as we do. And this makes it natural to conclude that if *we* are justified on the basis of this evidence, then demon-deceived subjects are equally justified in exactly the same sense.

However, I have been arguing against the experientialist thesis that perceptual experiences must serve as the evidence which justifies our perceptual beliefs, and have attempted to make plausible that our perceptual beliefs can be non-evidentially justified. And even if one is convinced that perceptual experiences do constitute evidence for our beliefs, then the Distinctiveness Problem (outlined in Chap. 3) should lead one to doubt that perceptual experiences are able to do so merely because they are experiences. Just as beliefs cannot evidentially justify beliefs merely because of their status as beliefs, so are experiences unable to evidentially justify beliefs merely because of their status as experiences. The etiology of the beliefs (are they justified themselves?) and experiences (are they reliably connected to the environment?) also matters. So even if one does not doubt that experiences *can* act as evidence for belief, then one should still doubt whether they can do so without being reliably connected to the environment. This means that even if Ned has evidence that is constituted by his perceptual experiences, then this evidence still falls short of being sufficient for justification, as it lacks a certain type of etiology that is important for their status as good evidence. On this perspective, Ned is like an epistemic agent who has random beliefs implanted in his mind: even though one might call these beliefs evidence of some sort, they certainly cannot evidentially justify any further beliefs. The best status that these further beliefs could achieve is that of being *conditionally* justified.[15]

If it is indeed true that the New Evil Demon Intuition is supported by a certain flawed view of experiences as evidence, then the New Evil Demon scenario cannot be used as an independent argument against a process reliabilist theory. The intuition would then presuppose an experientialist assumption that is explicitly rejected by process reliabilism, and for good reasons.

This explanation of the New Evil Demon Intuition assumes that we start from an incorrect theoretical view of perceptual justification, namely, as being supported purely by experiential evidence, and then go on to apply that view to radically deceived subjects. But one might wonder whether this is a fair way of dealing with the intuition. After all, the New Evil Demon scenario is supposed to elicit *pre-theoretical* intuitions about justification. Why should we suppose that these intuitions are based on a certain theoretical account of perceptual justification?

[14] Internalists that accept this assumption are mentioned in Chaps. 2 and 3, for externalists that accept the assumption see, e.g., Greco (2000), Comesaña (2010), and Goldman (2011b).

[15] See the previous chapter, Sect. 5.3.1 for more on this notion of conditional justification.

What is needed is an explanation of why we would, in making ordinary judgments about perceptual justification, mistakenly attach so much importance to perceptual experience.

Fortunately, such an explanation can be provided. First of all, in the actual world a perceptual experience that p is usually reliably connected to the fact that p. Although we are familiar with perceptual illusions, in general the world appears to be in accordance with our perceptual experiences of it. But then it is no wonder that we take experience to be an important factor in perceptual justification: perceptual experience, after all, is *actually* reliably connected to how the world is. Of course, the theoretical considerations I have put forward are meant to support the idea that, in fact, neither conscious experience nor actuality are the important factors for perceptual justification. But this is compatible with us having a mistaken view about the importance of such factors.

Second, when our perceptual beliefs are challenged ("how can you be sure that p?"), we often can and do answer by saying, e.g., "because I see that p". Epistemological disjunctivists have used this fact about our ordinary practices as a way to motivate their theory of perceptual justification, where this type of justification crucially depends on the fact that one perceives that such-and-so is the case. But one can equally understand these ordinary ways of bolstering one's perceptual claims as appealing to what one is currently *experiencing*, or to what one currently believes to be experiencing. Seen in this light, our practice of giving reasons for perceptual beliefs could reinforce the idea that conscious experience is *the* important justificatory factor when dealing with perceptual justification.

The actual reliability of perceptual experience, together with our practice of responding to challenges of our perception-based claims, can thus provide a tentative explanation of why we would attach too much epistemic weight to perceptual experience, thereby making us into natural experientialists. And if we are such natural experientialists, then it is easy to see why we would have the mistaken intuition that radically deceived subjects like Ned have justified perceptual beliefs.

We've now seen how a non-evidentialist about perceptual justification could explain our mistaken intuition about the New Evil Demon scenario. But it might not yet be clear to what such a non-evidentialist account of justification is committed exactly. The next section will be devoted to making these commitments more clear.

6.6 Non-evidential Perceptual Justification

Throughout this book my main aim has been to show the importance of accepting a non-evidential account of perceptual justification, which allows for perceptual justification even in the absence of perceptual experience. Because of this, I have not clearly taken a stand on the different varieties of reliabilist theories with which such an account would be compatible. However, I have committed myself to several ideas that I take to be crucial for any satisfactory theory of perceptual justification. What's more, I have attempted to show that further additions to this fairly lean version of process reliabilism are at least not motivated by considerations related to the case of perception. So what does such a lean version of reliabilism look like?

To begin with, reliabilists need to distinguish between evidential and non-evidential justification, where that can either be taken as two different types of relations or as two different species of properties that belong to the same genus (see Sect. 2.3). Evidential justification has to do with the contribution to justification made by a subject's reasons or evidence, whereas non-evidential justification has to do with the contribution to justification made by non-evidential factors. To be fully successful, a process reliabilist account has to spell out both these types of justification in terms of the (conditional) reliability of the cognitive processes involved. I have not provided such an account for evidential justification, but two plausible necessary conditions are that the evidence in virtue of which the evidential justification obtains is itself justified (be it evidentially or non-evidentially), and that the process from evidence to conclusion is conditionally reliable, i.e., that the process has a high truth-ratio given accurate inputs. In contrast, non-evidential justification requires that a belief is the output of a specific kind of reliable cognitive process (on the classic process reliabilist picture), or, if one prefers, the output of a properly functioning cognitive system that has the aim of reliably producing true beliefs (on the proper functionalist picture).

The notions of evidential and non-evidential justification are closely connected to those of epistemically basic and non-basic beliefs. Given that evidential justification already presupposes that the supporting evidence is justified, we need to have some foundational evidence that need not be evidentially justified itself. This foundation is constituted by basic beliefs, beliefs that are capable of being justified by just being the outcome of a reliable or properly functioning process or system. Now, we've seen that the notions of belief-dependent and belief-independent processes were not useful in distinguishing between basic and non-basic beliefs (Sect. 5.5). Even if a subject's cognitive process uses a certain belief of that subject as input, that does not mean that it thereby uses the content of that belief as evidence for its conclusion. Introspection is a case in point: in introspection the belief state rather than its content is used to output a higher-order belief about that state. So reliabilists need an alternative account of basic beliefs, and the one provided by Lyons (2009) certainly looks plausible once stripped of its etiological condition: basic beliefs are those that are produced by the non-inferential operation of an inferentially opaque cognitive system. Basic beliefs can then be taken to be prima facie justified if and only if they resulted from a reliable or properly functioning cognitive process.

Nowhere in this account of basic belief and justification does it say that basic beliefs cannot be further supported by means of evidential justification. Indeed, I have proposed that our basic perceptual beliefs typically *are* further supported by a specific kind of higher-order evidence. By appealing to a reliable introspective monitoring mechanism, reliabilists can easily account for justified higher-order beliefs that one is currently seeing that such-and-so is the case. These types of justified higher-order beliefs can then be used to strengthen the epistemic status of basic perceptual beliefs.

But reliabilists should also have something to say about the way in which the justification of basic beliefs can be defeated. It seems that the prima facie justification of a subject's perceptual belief is defeated if an available properly functioning monitoring mechanism would not have led the subject to that belief. A full-blown process reliabilist way of motivating such an addition could be found in the idea that ultima facie justification not only requires the reliability of the cognitive processes which produce the belief, but also the reliability of those cognitive processes which *sustain* the belief. If one extrapolates from this idea, one ends up with the general claim that a belief is defeated if it is sustained by a (conditionally) unreliable process. But the earlier counterfactual claim still remains important as claims about sustainability definitely appear connected to what would have happened under different conditions.

These are the key ingredients of the non-evidentialist view of perceptual justification that I have tried to make plausible. I will now turn to some objections that this account might evoke in defenders of experientialism or epistemological disjunctivism.

6.6.1 The Epistemic Role of Experience

One of the ingredients of the proposed non-evidential account of perceptual justification might be hard to swallow for experientialists. According to the account, basic beliefs are prima facie justified if and only if they are the output of a specific sort of process that need not include any sort of experiential component. What's more, defeat is also detachable from any experiential component as this merely has to do with the absence of an unreliable belief-sustaining mechanism. This means that the proposed non-evidential account not only allows that perceptual beliefs can be justified in the absence of experience, it also seems to allow that experience has *no* epistemic role to play whatsoever. Mark Johnston argues against this point as follows:

> Surely there is some distinctive positive epistemic virtue exemplified by the normally sighted, who arrive at the knowledge that there is a pineapple before them by seeing the pineapple before them, or more carefully, who have knowledge that there is a pineapple before them and see the pineapple there as well. Surely sight confers a distinctive epistemic advantage on us. As does audition, smell, taste and touch. Surely there is something per se epistemically defective about being insensate, even though your immediate perceptual beliefs are justified, reliably formed, and in fact often constitute knowledge.
>
> (Johnston 2011, p. 167)

However, there are several ways in which one can accommodate the thought that there is a positive epistemic virtue exemplified by the normally sighted without accepting that perceptual experience evidentially justifies belief. First of all, the proposed account stresses that the normally sighted, interpreted as normally sighted human beings, not only see that *p* but also know that they see that *p*. If one supposes that higher-order beliefs of this type are necessary and sufficient for the specific conscious experiences these normally sighted individuals undergo, then it follows that the normally sighted are in this respect epistemically better off than many types of 'zombies' (i.e., subjects who lack perceptual experiences but still have reliable perceptual faculties).[16]

Second, even though conscious experience might not play any *evidential* role, it could still contribute to justification by contributing to the overall reliability of the perceptual system. Perhaps our perceptual system would not be as reliable without the contribution of conscious experience, which would again make our epistemic situation preferable to that of zombies.

Third, there is also the possibility of allowing experience to play its evidential role in virtue of its having a certain etiology, such as being reliably connected to the environment. However, to provide a fully satisfying answer to Johnston's worry one would then also have to explain why these experiences would improve one's epistemic position over the position of zombies. It's not so clear how that could be done, especially once one considers that zombies could also evidentially support their beliefs by invoking higher-order beliefs about what they're perceiving.

Let me also mention another suggestion about the epistemic role of experience that is provided by Lyons (2009):

> My having a headache is relevant to the epistemic status of my belief that I don't have a headache, but not evidentially, as would be, for instance, the belief that I've just been hit in the head. Rather, it is significant in the way that the room's being conspicuously full of people is relevant to my belief that the room is empty. If I use induction to arrive at the belief that the room is empty, when it's conspicuously not, the use of vision or some other reliable process would result in my not believing that the room is empty [...] So my belief that the room is empty is prima facie justified in virtue of the reliability of induction, but the belief is defeated [...] The belief is (ultima facie) unjustified because there is an available alternative reliable process, which, if used in addition to the one actually used, would have resulted in my not believing the room was empty.
>
> (Lyons 2009, pp. 175–6)

According to Lyons, what is epistemically important about having perceptual experiences is that they are part of a "highly reliable" (2009, p. 176) and "highly available" (ibid.) alternative belief-forming process, viz., perception, which can defeat beliefs that arise out of non-perceptual grounds (one might say that they are non-evidential defeaters for beliefs). This also explains why Lyons discusses an example in which a room is *conspicuously* full of people: being conspicuously full of people just means that there is a highly *available* process that would lead to the belief that the room was full of people, namely, quick and easy vision.

Now, note that this proposal uses ARP as a condition on defeat, a condition that I have criticized above. But let's ignore that for a moment. The problem is

[16] I have presented a way in which higher-order beliefs could influence perceptual phenomenology in Sect. 6.3.1. A more radical position is provided by so-called higher-order theories of consciousness, according to which consciousness has to do with a subject's higher-order capabilities (e.g., higher-order perception (HOP) of one's first-order states or higher-order thought (HOT) about one's first-order states) (Gennaro 2004).

that philosophers with Johnston's worry will still remain unimpressed by such an account of the epistemic role of experience. There's nothing essential that connects this type of defeat with perceptual *experience* rather than an insensate perceptual process with the same functional characteristics. So, like the third point above, this does not seem to grant an epistemic role to perceptual experience specifically. In contrast, if perceptual experience enhances the reliability of our perceptual system, or if the phenomenology of perceptual experience is partly reducible to certain types of higher-order beliefs, then perceptual experience itself really has an important epistemic role to play.

6.6.2 The Accessibility Intuition

Another objection to the proposed non-evidentialist theory of perceptual justification might stem from the fact that it no longer adheres to the thought that justification requires accessible reasons. According to the proposed account, an epistemic agent can have justified perceptual beliefs without having any access to evidence or reasons for those beliefs. All that's necessary for perceptual justification is a specific sort of reliable process, which means that an agent need not even be aware of the fact that she is using such a reliable process. This seems to go against the idea that people should have accessible reasons for their beliefs.

In contrast, both experientialism and epistemological disjunctivism make sure that an agent only has justified perceptual beliefs if he also has access to reasons for those beliefs. Experientialism takes the reasons to be constituted by experience itself, while epistemological disjunctivism takes the reasons to be constituted by the fact that one is perceiving that such-and-so is the case. Doesn't this constitute a point in favor of experientialist and epistemological disjunctivist theories of justification?

In fact, I think that considerations about accessibility actually count in favor of the non-evidentialist theory. First, note that it's unclear in what sense unsophisticated agents are supposed to have accessible reasons for their beliefs. Even though we can assume that these agents are still able to undergo perceptual experiences, it's not clear that this is sufficient to have any real kind of possible access to a reason for belief. Second, for the case of sophisticated epistemic agents, I have argued that justified perceptual belief typically does come hand in hand with access to reasons. When we see that such-and-so is the case, we typically also know that we see that such-and-so is the case. What's more, this type of access can be accounted for in terms of introspective mechanisms that epistemically parallel perceptual mechanisms. The introspective mechanisms produce beliefs on the basis of cues that are largely unconscious, and these beliefs are justified because this process is also reliable. This account thus has the benefit of positing a type of access to reasons that is not left mysterious.

This means that, when it's unclear whether it actually makes sense to speak of access to reasons—as in the case of unsophisticated cognizers—the non-evidentialist theory can nevertheless account for justified belief, and when it does make sense to speak of access to reasons, the non-evidentialist theory can provide a worked out account of the type of access involved. On both these fronts it thus scores better than many evidentialist alternatives.

6.6.3 Introspective Mechanisms in Epistemological Disjunctivism

A final critical consideration has to do with the introspective mechanisms that are part of the proposed theory of perceptual justification. These introspective mechanisms are supposed to account for justified higher-order beliefs of the form "I see that *p*" without falling into the trap of hyper-intellectualization. It can do so because the introspective mechanisms are supposed to work without any conscious reflection on the part of the subject. But one of the major arguments against epistemological disjunctivism precisely exploited the worry about hyper-intellectualization (Chap. 4), so one might think that epistemological disjunctivism can be improved by combining it with the idea of unconscious introspective mechanisms.

In fact, the idea that higher-order recognitional mechanisms in an important sense parallel first-order perceptual-recognitional mechanisms is one of the crucial aspects of Millar's (2010) Knowledge First Epistemological Disjunctivism. But even if one stresses that the introspective mechanisms are so simple as to apply to the more sophisticated types of animals, there will still always be unsophisticated cognizers that lack them. As long as such cognizers have the necessary first-order perceptual mechanisms, they will be capable of forming perceptual beliefs. And given that epistemological disjunctivism, on this proposal, requires that these perceptual beliefs are backed up by higher-order beliefs outputted by reliable introspective mechanisms, these cognizers will simply lack perceptual justification. This means that the hyper-intellectualization worry will not be solved by appealing to unconscious introspective mechanisms.

What's more, introspective mechanisms would provide an account of introspective justification that would not parallel the epistemological disjunctivist account of perceptual justification. The idea is precisely that introspective mechanisms are capable of providing *non-evidential* justification for higher-order beliefs because of the fact that they work reliably. In contrast, epistemological disjunctivists propose that perceptual justification is *evidential* in nature. A theory of justification that can be applied to both introspection and perception in the same way should surely be preferred. So epistemological disjunctivism simply does not fit right with a process reliabilist account of introspection.

6.7 Conclusion

In this chapter I have defended an alternative way to deal with the Clairvoyance and New Evil Demon Problem in a process reliabilist framework. The starting point for this alternative account comes from the epistemological disjunctivist's insight that sophisticated agents have a specific type of evidential justification at their disposal that stems from their higher-order perspective. I have extended this explanation in a process reliabilist way by showing how an introspective belief that one perceives that *p* can be justified by a reliable introspective process. It's plausible that our introspective processes make extensive use of unconscious and even non-accessible cues to output beliefs about our mental states, which makes it fit nicely with an externalist rather than internalist theory of justification. In fact, our perceptual phenomenology might be best explained as being the *output* of unconscious introspective processes rather than that this perceptual phenomenology constitutes the *input* to such processes.

Moreover, I have added the thought that higher-order processes are not only capable of providing evidential justification for first-order beliefs, but also capable of defeating those first-order beliefs by the way they monitor first-order belief-forming processes. The best explanation of this type of defeat seems to come from a proper functionalist perspective: if a subject's monitoring mechanism would have prevented a belief from occurring if it had been functioning properly, then this belief is defeated. This is what goes on in cases of clairvoyance: although the respective clairvoyants all have beliefs that are formed by reliable processes, they also have defeater systems in the form of monitoring mechanisms that should have kept them from accepting those beliefs on the basis of the content and context of the incoming information.

The explanation of the New Evil Demon Intuition also partly derives from the idea that we usually have higher-order reasons for our perceptual beliefs. Because we usually appeal to the fact that we see that *p* in our justification for believing that *p*, we tend to assign an evidential role to experience that it does not in fact have. This experientialist view of experience is even exacerbated by the fact that experience actually is reliably connected to our environment. It is therefore no surprise that victims of an evil demon seem to have justified perceptual beliefs because they have the same experiences as us. However, once we realize that experiences need not play any evidential role whatsoever, or at least are unable to play that evidential role merely in virtue of their status as experiences, then we should no longer be too impressed by the New Evil Demon Intuition.

These considerations lead me to conclude that we should uphold a non-evidential theory of perceptual justification. By allowing for the possibility of non-evidential perceptual justification and evidential justification by higher-order reasons, we can explain both how unsophisticated and sophisticated agents are able to have justified perceptual beliefs without falling into the trap of either hyper-intellectualization or assimilating sophisticated justification to unsophisticated justification. What's more, this theory also brings with it a nice account of access for higher-order reasons in the form of a reliable introspective mechanism. The only thing the theory doesn't obviously do is provide a necessary epistemic role for conscious experience, but, if I'm correct, this role is overrated anyway.

Bibliography

Alston, W. P. (1988). An internalist externalism. *Synthese, 74*(3), 265–283.
Alston, W. P. (1995). How to think about reliability. *Philosophical Topics, 23*(1), 1–29.
Barth, H., Kanwisher, N., & Spelke, E. (2003). The construction of large number representations in adults. *Cognition, 86*(3), 201–221.
Bayne, T. (2009). Perception and the reach of phenomenal content. *The Philosophical Quarterly, 59*(236), 385–404.
Bayne, T., & Spener, M. (2010). Introspective humility. *Philosophical Issues, 20*(1), 1–22.
Beddor, B. (2015). Process reliabilism's troubles with defeat. *The Philosophical Quarterly, 65*(259), 145–159.
Bedke, M. S. (2010). Developmental process reliabilism: On justification, defeat, and evidence. *Erkenntnis, 73*(1), 1–17.
Beebe, J. R. (2004). Reliabilism, truetemp and new perceptual faculties. *Synthese, 140*(3), 307–329.
Bergmann, M. (2006). *Justification without awareness: A defense of epistemic externalism*. Oxford Scholarship Online. Philosophy module. Oxford: Clarendon Press.
Bergmann, M. (2013). Externalist justification and the role of seemings. *Philosophical Studies, 166*, 163–184.
Bernecker, S. (2008). Agent reliabilism and the problem of clairvoyance. *Philosophy and Phenomenological Research, 76*(1), 164–172.
Bird, A. (2007). Justified judging. *Philosophy and Phenomenological Research, 74*(1), 81–110.
BonJour, L. (1985). *The structure of empirical knowledge*. Cambridge: Harvard University Press.
BonJour, L., & Sosa, E. (2003). *Epistemic justification: Internalism Vs. externalism, foundations Vs. virtues*. Malden: Blackwell Publishing.
Breyer, D., & Greco, J. (2008). Cognitive integration and the ownership of belief: Response to bernecker. *Philosophy and Phenomenological Research, 76*(1), 173–184.
Brogaard, B. (2013a). Do we perceive natural kind properties? *Philosophical Studies, 162*(1), 35–42.
Brogaard, B. (2013b). Phenomenal seemings and sensible dogmatism. In C. Tucker (Ed.), *Seemings and justification: New essays on dogmatism and phenomenal conservatism* (pp. 270–289). New York: Oxford University Press.
Burge, T. (2003). Perceptual entitlement. *Philosophy and Phenomenological Research, 67*(3), 503–548.
Burge, T. (2010). *Origins of objectivity*. New York: Oxford University Press.
Carruthers, P. (2011). *The opacity of mind: An integrative theory of self-knowledge*. New York: Oxford University Press.

H. Ghijsen, *The Puzzle of Perceptual Justification*, Synthese Library 377,
DOI 10.1007/978-3-319-30500-4

Cassam, Q. (2007). *The possibility of knowledge*. Oxford: Oxford University Press.
Chisholm, R. (1977). *Theory of knowledge* (2nd ed.). Englewood Cliffs: Prentice-Hall.
Chudnoff, E. (2011). The nature of intuitive justification. *Philosophical Studies, 153*, 313–333.
Chudnoff, E. (2012). Presentational phenomenology. In S. Miguens & G. Preyer (Eds.), *Consciousness and subjectivity* (pp. 51–72). Frankfurt/New Brunswick: Ontos Verlag.
Chudnoff, E. (2013). *Intuition*. New York: Oxford University Press.
Chudnoff, E. (2014). Review of chris tucker's (ed.) *Seemings and justification: New essays on dogmatism and phenomenal conservatism. Notre Dame Philosophical Reviews*. Notre Dame: University of Notre Dame. http://ndpr.nd.edu/news/48118-seemings-and-justification-new-essays-on-dogmatism-and-phenomenal-conservatism/
Comesaña, J. (2010). Evidentialist reliabilism. *Noûs, 44*(4), 571–600.
Conee, E., & Feldman, R. (1998). The generality problem for reliabilism. *Philosophical Studies, 89*(1), 1–29.
Conee, E., & Feldman, R. (2004a). *Evidentialism: Essays in epistemology*. New York: Clarendon Press.
Conee, E., & Feldman, R. (2004b). Making sense of skepticism. In E. Conee & R. Feldman (Eds.), *Evidentialism: Essays in epistemology* (pp. 277–306). New York: Clarendon Press.
Cowey, A. (2010). The blindsight saga. *Experimental brain research, 200*(1), 3–24.
Crane, T. (2011). The problem of perception. In E. N. Zalta (Ed.), *The Stanford encyclopedia of philosophy* (Spring 2011 ed.). Stanford University, Stanford University http://plato.stanford.edu/archives/spr2011/entries/perception-problem/
Dennett, D. C. (1991). *Consciousness explained*. New York: Back Bay Books.
DePaul, M. (2009). Phenomenal conservatism and self-defeat. *Philosophy and Phenomenological Research, 78*(1), 205–212.
Dokic, J. (2012). Seeds of self-knowledge: Noetic feelings and metacognition. In M. J. Beran, J. Brandl, J. Perner, & J. Proust (Eds.), *Foundations of metacognition* (pp. 302–321). Oxford: Oxford University Press.
Dretske, F. (1981). *Knowledge and the flow of information*. Cambridge: MIT.
Feldman, R. (1985). Reliability and justification. *The Monist, 68*(2), 159–174.
Feldman, R., & Conee, E. (1985). Evidentialism. *Philosophical Studies, 48*(1), 15–34.
Feldman, R., & Conee, E. (2001). Internalism defended. *American Philosophical Quarterly, 38*(1), 1–18.
Feldman, R., & Conee, E. (2002). Typing problems. *Philosophy and Phenomenological Research, 65*(1), 98–105.
Fish, W. (2008). Disjunctivism, indistinguishability, and the nature of hallucination. In A. Haddock & F. Macpherson (Eds.), *Disjunctivism: Perception, action, knowledge*, (pp. 144–167). Oxford: Oxford University Press.
Fish, W. (2009). *Perception, hallucination, and illusion*. New York: Oxford University Press.
Fish, W. (2013). High-level properties and visual experience. *Philosophical Studies, 162*(1), 43–55.
French, C. (2012). Does propositional seeing entail propositional knowledge? *Theoria, 78*, 115–127.
Fumerton, R. (1995). *Metaepistemology and skepticism*. Lanham: Rowman & Littlefield.
Gennaro, R. J. (2004). *Higher-order theories of consciousness: An anthology*. Amsterdam: John Benjamins Publishing.
Ghijsen, H. (2015). The real epistemic problem of cognitive penetration (online first). *Philosophical Studies*. doi: 10.1007/s11098-015-0558-2.
Goldman, A. (1976). Discrimination and perceptual knowledge. *The Journal of Philosophy, 73*(20), 771–791.
Goldman, A. (1979). What is justified belief? In G. Pappas (Ed.), *Justification and knowledge* (pp. 1–23). Dordrecht: D. Reidel.
Goldman, A. (1986). *Epistemology and cognition*. Cambridge: Harvard University Press.
Goldman, A. (1988). Strong and weak justification. *Philosophical Perspectives, 2*, 51–69.
Goldman, A. (2011a). Reliabilism. In E. N. Zalta (Ed.), *The Stanford encyclopedia of philosophy* (Spring 2011 ed.). http://plato.stanford.edu/archives/spr2011/entries/reliabilism/.

Goldman, A. (2011b). Toward a synthesis of reliabilism and evidentialism? or: Evidentialism's troubles, reliabilism's rescue package. In T. Dougherty (Ed.), *Evidentialism and its discontents* (pp. 254–280). New York: Oxford University Press.

Goldman, A. I. (1967). A causal theory of knowing. *The Journal of Philosophy, 64*(12), 357–372.

Goldman, A. I. (2006). *Simulating minds: The philosophy, psychology, and neuroscience of mindreading*. New York: Oxford University Press.

Goldman, A. I. (2009). Replies to discussants. *Grazer Philosophische Studien, 79*(1), 245–288.

Graham, P. J. (2011). Perceptual entitlement and basic beliefs. *Philosophical Studies, 153*(3), 467–475.

Graham, P. J. (2012). Epistemic entitlement. *Noûs, 46*(3), 449–482.

Graham, P. J. (2014). Functions, warrant, history. In A. Fairweather & O. Flanagan (Eds.), *Naturalizing epistemic virtue* (pp. 15–35). New York: Cambridge University Press.

Greco, J. (2000). *Putting skeptics in their place: The nature of skeptical arguments and their role in philosophical inquiry* (2007 digital print ed.). New York: Cambridge University Press

Greco, J. (2003). Further thoughts on agent reliabilism: Replies to cohen, geivett, kvanvig, and schmitt and lahroodi. *Philosophy and Phenomenological Research, 66*(2), 466–480.

Grundmann, T. (2009). Reliabilism and the problem of defeaters. *Grazer Philosophische Studien, 79*(1), 65–76.

Hansen, T., Olkkonen, M., Walter, S., & Gegenfurtner, K. R. (2006). Memory modulates color appearance. *Nature Neuroscience, 9*(11), 1367–1368.

Horgan, T. (2012). Introspection about phenomenal consciousness: Running the gamut from infallibility to impotence. In D. Smithies & D. Stoljar (Eds.), *Introspection and consciousness* (pp. 405–422). New York: Oxford University Press.

Horgan, T., & Kriegel, U. (2007). Phenomenal epistemology: What is consciousness that we may know it so well? *Philosophical Issues, 17*(1), 123–144.

Huemer, M. (2001). *Skepticism and the veil of perception*. Lanham: Rowman & Littlefield Publishers.

Huemer, M. (2007). Compassionate phenomenal conservatism. *Philosophy and Phenomenological Research, 74*(1), 30–55.

Huemer, M. (2013). Epistemological asymmetries between belief and experience. *Philosophical Studies, 162*(3), 741–748.

Hurlburt, R. T., & Schwitzgebel, E. (2007). *Describing inner experience? Proponent meets skeptic*. Cambridge: MIT.

Johnson, D. M. (2011). Proper function and defeating experiences. *Synthese, 182*(3), 433–447.

Johnson, M., Hashtroudi, S., & Lindsay, D. (1993). Source monitoring. *Psychological Bulletin, 114*(1), 3–28.

Johnston, M. (2011). On a neglected epistemic virtue. *Philosophical Issues, 21*(1), 165–218.

Kelp, C. (2011). Not without justification. *Dialectica, 65*(4), 581–595.

Klein, P. (1998). Foundationalism and the infinite regress of reasons. *Philosophy and Phenomenological Research, 58*, 919–925.

Kornblith, H. (2008). Knowledge needs no justification. In Q. Smith (Ed.) *Epistemology: New essays* (pp. 5–23). New York: Oxford University Press.

Kornblith, H. (2015). The role of reasons in epistemology. *Episteme, 12*(2), 225–239.

Kvanvig, J. (2007). Two approaches to epistemic defeat. In D.-P. Baker (Ed.), *Alvin plantinga* (pp. 107–124). New York: Cambridge University Press.

Lehrer, K. (1990). *Theory of knowledge*. Boulder: Westview Press.

Lehrer, K., & Cohen, S. (1983). Justification, truth, and coherence. *Synthese, 55*(2), 191–207.

Lyons, J. (2009). *Perception and basic beliefs: Zombies, modules and the problem of the external world*. USA: Oxford University Press.

Lyons, J. (2011). Circularity, reliability, and the cognitive penetrability of perception. *Philosophical Issues, 21*(1), 289–311.

Lyons, J. C. (2013). Should reliabilists be worried about demon worlds? *Philosophy and Phenomenological Research, 86*(1), 1–40.

Madison, B. (2013). Epistemological disjunctivism and the new evil demon. *Acta Analytica*, 1–10. Online first. doi:10.1007/s12136-013-0194-4.

Majors, B., & Sawyer, S. (2005). The epistemological argument for content externalism. *Philosophical Perspectives, 19*(1), 257–280.

Markie, P. (2005). The mystery of direct perceptual justification. *Philosophical Studies, 126*(3), 347–373.

Martin, M. (2006). On being alienated. In T. S. Gendler & J. Hawthorne (Eds.), *Perceptual experience* (pp. 354–410). Oxford: Clarendon Press.

Masrour, F. (2011). Is perceptual phenomenology thin? *Philosophy and Phenomenological Research, 83*(2), 366–397.

McDowell, J. (1982). Criteria, defeasibility, and knowledge. *Proceedings of the British Academy, 68*(1981), 455–479. Reprinted in Byrne and Logue. (2009). *Disjunctivism: Contemporary readings* (pp. 75–90). Cambridge: MIT.

McDowell, J. (1994). *Mind and world: With a new introduction* (1996 ed.). Cambridge: Harvard University Press.

McDowell, J. (2010). Tyler burge on disjunctivism. *Philosophical Explorations, 13*(3), 243–255.

McDowell, J. (2013). Tyler burge on disjunctivism (ii). *Philosophical Explorations, 16*(3), 259–279.

McGrath, M. (2013a). Phenomenal conservatism and cognitive penetration: The "bad basis" counterexamples. In C. Tucker (Ed.), *Seemings and justification* (pp. 225–247). New York: Oxford University Press.

McGrath, M. (2013b). Siegel and the impact for epistemological internalism. *Philosophical Studies, 162*(3), 723–732.

Millar, A. (2010). Knowledge and recognition. In D. Pritchard, A. Millar, & A. Haddock (Eds.), *The nature and value of knowledge: Three investigations* (pp. 89–188). Oxford: Oxford University Press.

Millar, A. (2011). How visual perception yields reasons for belief. *Philosophical Issues, 21*(1), 332–351.

Millikan, R. G. (1989). Biosemantics. *The Journal of Philosophy, 86*(6), 281–297.

Nagel, T. (1974). What is it like to be a bat? *The Philosophical Review, 83*(4), 435–450.

Nichols, S., & Stich, S. (2003). *Mindreading: An integrated account of pretence, self-awareness, and understanding other minds*. New York: Oxford University Press.

Overgaard, M. (2011). Visual experience and blindsight: A methodological review. *Experimental Brain Research, 209*, 473–479.

Pascual-Leone, A., & Walsh, V. (2001). Fast backprojections from the motion to the primary visual area necessary for visual awareness. *Science, 292*(5516), 510–512.

Perky, C. (1910). An experimental study of imagination. *The American Journal of Psychology, 21*(3), 422–452.

Plantinga, A. (1993). *Warrant and proper function*. New York: Oxford University Press.

Plantinga, A. (1996). Alvin plantinga replies. In J. Kvanvig (Ed.) *Warrant in contemporary epistemology: Essays in honor of Plantinga's theory of knowledge* (pp. 307–378). Lanham: Rowman & Littlefield Publishers, Inc.

Plantinga, A. (2000). *Warranted Christian belief*. New York: Oxford University Press.

Pollock, J. (1974). *Knowledge and justification*. Princeton: Princeton University Press.

Price, H. (1932). *Perception* (1964 ed.). London: Methuen & Co. Ltd.

Prinz, J. (2006). Beyond appearances: The content of sensation and perception. In T. S. Gendler & J. Hawthorne (Eds.), *Perceptual experience* (pp. 434–461). New York: Oxford University Press.

Pritchard, D. (2006). *What is this thing called knowledge?* (2010 ed.). New York: Routledge.

Pritchard, D. (2008). Mcdowellian neo-mooreanism. In A. Haddock & F. Macpherson (Eds.), *Disjunctivism: Perception, action, knowledge* (pp. 283–310). Oxford: Oxford University Press.

Pritchard, D. (2012a). Anti-luck virtue epistemology. *The Journal of Philosophy, 109*(3), 247–279.

Pritchard, D. (2012b). *Epistemological disjunctivism*. Oxford: Oxford University Press.

Pryor, J. (2000). The skeptic and the dogmatist. *Noûs, 34*(4), 517–549.

Pryor, J. (2004). What's wrong with moore's argument? *Philosophical Issues, 14*, 349–378.

Pryor, J. (2005). There is immediate justification. In M. Steup & E. Sosa (Eds.), *Contemporary debates in epistemology* (pp. 181–201). Malden: Blackwell Publishing.

Putnam, H. (1973). Meaning and reference. *The Journal of Philosophy, 70*(19), 699–711.

Reeves, A., & Craver-Lemley, C. (2012). Unmasking the perky effect: Spatial extent of image interference on visual acuity. *Frontiers in Psychology, 3*(296), 1–7.

Reiland, I. (2014). On experiencing high-level properties. *American Philosophical Quarterly, 51*(3), 177–187.

Robinson, H. (1994). *Perception*. London/New York: Routledge.

Ruloff, C. (2000). Some remarks on bonjour on warrant, proper function, and defeasibility. *Principia: An International Journal of Epistemology, 4*(2), 215–228.

Schwitzgebel, E. (2008). The unreliability of naive introspection. *Philosophical Review, 117*(2), 245–273.

Schwitzgebel, E. (2012). Introspection, what? In D. Smithies & D. Stoljar (Eds.), *Introspection and consciousness* (pp. 29–48). New York: Oxford University Press.

Segal, S., & Gordon, P. (1969). The perky effect revisited: Blocking of visual signals by imagery. *Perceptual and Motor Skills, 28*(3), 791–797.

Sellars, W. (1956). *Empiricism and the philosophy of mind, intro. by Richard Rorty and study guide by Robert Brandom* (1997 ed.). Cambridge: Harvard University Press.

Siegel, S. (2006). Which properties are represented in perception? In T. S. Gendler & J. Hawthorne (Eds.), *Perceptual experience* (pp. 481–503). New York: Oxford University Press.

Siegel, S. (2012). Cognitive penetrability and perceptual justification. *Noûs, 46*, 201–222.

Siegel, S. (2013a). The epistemic impact of the etiology of experience. *Philosophical Studies, 162*(3), 697–722.

Siegel, S. (2013b). Reply to fumerton, huemer, and mcgrath. *Philosophical Studies, 162*(3), 749–757.

Silvanto, J., Cowey, A., Lavie, N., & Walsh, V. (2005). Striate cortex (v1) activity gates awareness of motion. *Nature Neuroscience, 8*(2), 143–144.

Simion, M. (2016). Perception, history and benefit. *Episteme, 13*(1), 61–76.

Smith, A. (2002). *The problem of perception*. Cambridge: Harvard University Press.

Smithies, D. (2012a). Mentalism and epistemic transparency. *Australasian Journal of Philosophy, 90*(4), 723–741.

Smithies, D. (2012b). A simple theory of introspection. In D. Smithies & D. Stoljar (Eds.), *Introspection and consciousness* (pp. 259–293). New York: Oxford University Press.

Smithies, D. (2013). Review of duncan pritchard's. In D. Pritchard (Ed.), *Epistemological Disjunctivism. Notre Dame philosophical reviews.* Notre Dame: University of Notre Dame. http://ndpr.nd.edu/news/36590-epistemological-disjunctivism

Sosa, E. (1993). Proper functionalism and virtue epistemology. *Noûs, 27*, 51–65.

Sosa, E. (2007). *A virtue epistemology: Apt belief and reflective knowledge* (Vol. I). Oxford: Clarendon Press.

Sosa, E. (2011). *Knowing full well*. Princeton: Princeton University Press.

Sosa, E. (2015). *Judgment and agency*. New York: Oxford University Press.

Steup, M. (1999). A defense of internalism. In L. Pojman (Ed.), *The theory of knowledge: Classical and contemporary readings* (2nd ed., pp. 373–384). Belmont: Wadsworth Publishing.

Tolhurst, W. (1998). Seemings. *American Philosophical Quarterly, 35*(3), 293–302.

Tucker, C. (2010). Why open-minded people should endorse dogmatism. *Philosophical Perspectives, 24*(1), 529–545.

Tucker, C. (2014). If dogmatists have a problem with cognitive penetration, you do too. *Dialectica, 68*(1), 35–62.

Tye, M. (2000). *Consciousness, color, and content*. Cambridge: MIT.

Tye, M. (2009). A new look at the speckled hen. *Analysis, 69*(2), 258–263.

Williamson, T. (1990). *Identity and discrimination, revised edition*. Oxford: Basil Blackwell.

Williamson, T. (2000). *Knowledge and its Limits*. Oxford: Oxford University Press.

Williamson, T. (2013). *Identity and discrimination* (Revised ed.). Hoboken: Wiley.